P. T. MARSHALL AND
G. M. HUGHES

THE PHYSIOLOGY OF MAMMALS AND OTHER VERTEBRATES

A TEXT-BOOK FOR SCHOOLS AND COLLEGES

CAMBRIDGE

AT THE UNIVERSITY PRESS · 1972

Published by the Syndics of the Cambridge University Press
Bentley House, 200 Euston Road, London NW1 2DB
American Branch: 32 East 57th Street, New York, N.Y.10022

Library of Congress Catalogue Card Number: 64–21557

ISBNs:
0 521 05678 0 Clothbound
0 521 09451 8 Paperback

First published 1965
Reprinted with corrections 1967
First paper back edition 1967
Reprinted 1972

Printed in Great Britain
at the University Printing House, Cambridge
(Brooke Crutchley, University Printer)

THE PHYSIOLOGY OF MAMMALS
AND OTHER VERTEBRATES

CONTENTS

8 Nervous coordination

9 The endocrine system

PREFACE

Biology is a very large and varied subject which may be subdivided in many different ways. A common and usual one is to consider living organisms at a series of different levels, beginning with whole populations, then at the individual, organ system, tissue, cellular and molecular levels. Throughout the history of biology there have been changes in the particular level which has received most study and also shifting fashions in the approach to a given or to several levels which were in vogue at that particular time. Often these fashions can be related to developments of new techniques which require the repetition and interpretation of previous work. Some aspects of the biological approach remain constant despite these winds of change and one of these is the relation between structure and function. This relationship can be discussed at all levels of organisation and it is basic to the approach given in this book.

A great deal of this approach tends to be at the organ system level and as such continues to present problems to the biologist, but at the present time there is a great deal of emphasis at a molecular level so that no modern functional approach to the subject would be complete without some inclusion of the biochemistry of cellular activities. In this field we try to present a brief account of the rapidly expanding aspects in the context of more classical biology and to emphasise some of the principal biochemical processes rather than give a detailed account of metabolic pathways. Here, as well as elsewhere in the book, space has not been sufficient to allow a critical approach, and while much of the anatomical and physiological material is now well established the same is not necessarily true of the most recent biochemical work.

Despite the interest and importance of cellular function much of it is hardly suitable for teaching or demonstration to elementary classes and it is the physiological approach to the vertebrates that forms, and is likely to form, the bulk of first courses in animal biology. It is the experience of the authors and many others in teaching biology to sixth-formers and students at university that few recent textbooks have attempted to summarise in an elementary way the vast knowledge gained by mammalian physiologists. Although basically this is a textbook of physiology it differs from most standard texts in that it has not been written primarily for medical students. Because of this, much comparative material, both anatomical and physiological, has been included. Relatively large amounts of anatomical material are included in order to emphasise to the student the importance of considering form and function together and not in isolation from one another. Furthermore, comparative material has been included to show the need for further investigation in this sort of study, both for its own sake and also because of the light it may shed on the functioning of mammals. The value of close understanding between comparative physiologists, mammalian physiologists and clinical physiologists is apparent at the research level at the present time and perhaps, by emphasizing this in the early training of all three types of student, we may hope to encourage such co-operation further.

The presentation of such an integrated approach abounds with problems and we are aware that what is given here contains many faults both in detail and in its general attitudes. It is, however, because we believe there is a great need for integration at this level of teaching that we have thought such an attempt to be worth while. We also know that there are many others who are far more qualified to write a book of this sort than ourselves and hope that if any of them should read our attempt they will be good enough to let us know where they think we have made errors. Some of the information has been presented in a diagrammatic way which has inevitably required a great deal of simplification. We only hope that the simplifications that we have made and the selection of data presented will not give rise to any fundamental misconceptions at this elementary stage of teaching.

In summary then, we hope to have shown the relevance of the study of vertebrates in the A level syllabuses to the potential medical student or biologist. The major object of the book is to present data in a way which will prepare the sixth-former for the type of functional approach he will have at the university.

Because of our awareness of the great breadth of the field that is covered in this book we have sought advice from many people whom we should like to thank. First of all, we should particularly like to thank Dr George Salt for suggesting the co-operation between ourselves, and for his constant advice during the production of this book. We are grateful to Dr W. E. Balfour of the Physiological Laboratory, Cambridge, for reading through the whole typescript. Individual chapters have been read by several of our friends, including that on the endocrine system by the late Dr H.E.Tunnicliffe;

that on disease by Dr F. E. Russell; on excretion by Dr A. P. M. Lockwood; and on the nervous system by Dr B. M. H. Bush. Much of the biochemical work was read critically by Dr R. Gregory of the Biochemistry Department. The diagrams of the cell and mitochondrion were devised by Dr A. V. Grimstone. We also wish to thank Mr B. Roberts for his helpful comments on the proof.

Throughout the many problems that have arisen during publication we have had much help from the editorial staff of the Cambridge University Press, to whom we would like to express our thanks.

We believe that an important feature of the book is the original drawings of histological and skeletal material made available by several laboratories, including Anatomy, Physiology and Zoology. The drawings were done by T. W. Armstrong, while still a pupil at The Leys School, and to him we would like to express our thanks.

G. M. H.
P. T. M.

August 1964

ACKNOWLEDGEMENTS

We would like to express our thanks to the following for permission to reproduce or modify their material:

J. and A. Churchill Ltd for Fig. 66, from Winton and Bayliss, *Human Physiology*, and the same publishers and Professor M. de B. Daly for Fig. 42 from *The Principles of Human Physiology*, 13th edn; G. Harrap and Co. Ltd for Fig. 41, after R. B. Whellock, *General Biology*; William Heinemann Ltd for Figs. 22, 25, 28, 29, 31, 32, 33 and 51, from Dr G. M. Hughes, *Comparative Physiology of Vertebrate Respiration*; also for Fig. 27, from Dr P. Clegg and A. G. Clegg, *Biology of the Mammal*; and the same publishers and Dr A. P. M. Lockwood for Fig. 69, from *Animal Fluids and Their Regulation*; Livingstone Ltd and Dr G. S. Dawes for Fig. 47*d*, based on his figure in the 13th edn of *The Textbook of Physiology and Biochemistry* by Bell, Davidson and Scarborough; Macmillan & Co. Ltd. for Fig. 8, from *Human Anatomy* by Hamilton, Boyd and Mossman; Oxford University Press and Professor J. Z. Young for Figs. 43, 72, 73, 85, 95 and 129, from *The Life of Mammals*; John Wiley and Sons Inc. for Fig. 149, from Torrey, *Morphogenesis of Vertebrates.*

We also wish to thank the Scientific Research Society of America, publishers of the *American Scientist*, for Fig. 134 (after Bargmann); Sir James Gray for Figs. 45, 46 and 87, based on his laboratory diagrams; Lord Adrian for Fig. 105 from *The Basis of Sensation*; Professor J. E. Harris for Fig. 91, based on 'The Role of the Fins in Swimming Fishes' (*J. Exp. Biol.* vol. 13); Dr P. A. Merton for Figs. 106 and 119, based on his article in the *British Medical Bulletin* (vol. 12); and Professor H. W. Mossman for Fig. 48 from 'Comparative Morphogenesis of the Fetal Membranes and Accessory Uterine Structures', *Contributions to Embryology*, vol. 26, from the Carnegie Institute, Washington. Finally we would like to thank the Company of Biologists for Fig. 94, taken from Dr O. R. Barclay's article on 'Amphibian Locomotion' (*J. Exp. Biol.* vol. 23).

INTRODUCTION

The word *physiology* was originally used by Jean Fernel as long ago as 1542 when its meaning was 'natural knowledge'. About the end of the nineteenth century it came into more general use to cover that branch of biology concerned with the function of living organisms. In general the study of function is aimed at determining the specific role of given organ systems and the conditions under which they are able to proceed and enable the organism to maintain itself in its environment. Inevitably physiological studies began on higher organisms, particularly mammals, because of the interest in the functioning of the human body. Despite many advances in the study of lower organisms, particularly during the last thirty years, the great bulk of our physiological knowledge continues to be based upon the investigation of mammals. Even some elementary aspects of the physiology of lower vertebrates such as the volumes and rates of flow in the cardiovascular system, remain almost completely un-investigated. The stress on the functioning of given organ systems in physiology sometimes gives rise to the misconception that each of these is considered in isolation and that analysis is the sole process whereby the physiologist proceeds. In fact an equally important aspect of the physiological approach is to study the interaction and integration between different organ systems and the means whereby the whole organism maintains itself under favourable and adverse environmental conditions.

The way in which the various organ systems become integrated to maintain the body in some degree of steady state has been recognised for some time and led W. B. Cannon (1932), in his important book *The Wisdom of the Body*, to recognise these phenomena under the title of *homeostasis*. Homeostasis therefore includes the coordinated physiological processes which maintain most of the steady states in living organisms. Similar general principles probably apply to the establishment, regulation and control of steady states for all levels of organisation from individual cells to whole populations. Homeostasis does not imply a static condition, for a given state fluctuates about a mean as the various controlling factors interact with one another. Furthermore the steady states to which the regulatory mechanisms are directed may shift with time but throughout remain under close control. A similar but more limited concept had earlier been used by Claude Bernard (1813–78) when he issued his famous dictum 'La fixité du milieu intérieur est la condition de la vie libre'. This ability is most highly developed in mammals where the temperature, water content of the body, tension of respiratory gases in the blood, circulation, blood pH and sugar content, etc., are all maintained within narrow limits. Fluctuations which can be frequently observed in certain physiological events, such as heart and respiratory rate, are produced in order to maintain constant some more important parameter of the system. For example, in respiration this is usually the CO_2 tension of the blood circulating to the respiratory centres of the brain.

Descriptions of the mechanisms of homeostasis have recently been clarified by the use of terms such as *feedback mechanisms*, borrowed from physical sciences. Feedback implies the regulation of the input to a given system by some part of the output. In most cases the feedback is negative in character so that it tends to stabilise the input and maintain the system at a given level. Should the feedback be positive it will tend to increase the change and the system may go into oscillation. Such oscillations may occur if the normal mechanisms become interfered with either experimentally or in disease. The analysis of feedback mechanisms by electrical engineers has led to the recognition of many quantitative approaches which are now being applied to the physiological systems of organisms, particularly the nervous system.

Homeostasis thus forms a valuable concept for integrating the activities of the body as a whole, but from a practical point of view it is necessary to describe the different physiological systems in isolation from one another, although indicating various points at which they interact; and this of course becomes particularly clear when we consider the coordination achieved by the nervous and endocrine systems.

The processes of physiology

In some ways living organisms such as mammals are like internal-combustion engines. Such a comparison immediately emphasises the essential physicochemical nature of the approach of a physiologist to living matter, for although he may recognise the possibility that other properties exist he tends to regard them as vitalistic and, from a practical point of view, not able to be investigated.

Both the living and non-living machine have similar essentials for their function which are as follows:

(*a*) A supply of energy-providing substance—fuel in the one case, food in the other (*nutrition*);

(*b*) a means of breaking down the substance so that its energy is made available. An engine does this by burning petrol mixed with air so rapidly that the resulting explosion drives the piston down the cylinder; the living organism on the other hand uses a series of chemical stages, each one yielding a small quantity of energy (*respiration*);

(*c*) a means of converting this energy into a type of work or activity appropriate to the machine or organism, that is, the transmission mechanism using a flywheel and gear system, or the ways in which energy-rich substances become coupled to different effector systems (*movement*);

(*d*) a means, or several, of conveying the substances concerned in the energy reaction to the place where it takes place, and a means of removal of the waste products of the reaction. In the engine this is the petrol pump and the waste products come out in the exhaust, whereas in the living organism we have *circulation* and *excretion*;

(*e*) a system for coordinating the activity of the machine or organism which will ensure its continued function in a way that is adjusted to given conditions. Ignition, steering, accelerator, valves, etc., in the engine, and the nervous and endocrine systems of the organism (*coordination*).

The study of physiology involves the study of these processes plus two more, *growth* and *reproduction*, which are characteristic of living, as distinct from non-living, systems. It will be worth while examining some of the points raised above in more detail.

The food

In the living machine the energy exchanges are complicated by the fact that the framework itself is continuously being replaced. This process also makes demands on the stores of energy; so fuel is being used to build the machine, to enable it to grow and to keep it maintained. Living organisms also need small quantities of essential substances, such as mineral salts and certain organic chemicals, which are not used to provide energy but are involved in the structure or functioning of the organism.

The name given to a substance utilised in this way by the body is a respiratory *substrate*. Unlike the internal-combustion engine, and more like the steam-engine, the body can use a great range of fuels or substrates from which energy can be obtained.

The most usual energy-providing substrate for living organisms, including the mammals, is glucose, a sugar containing carbon, hydrogen and oxygen in the molecule and belonging to a class of compounds called *carbohydrates*. A number of other carbon–hydrogen–oxygen compounds can also be broken down to release energy, including the large-moleculed fatty acids as well as smaller compounds such as alcohol (C_2H_5OH).

Thus the respiratory, or energy-releasing, mechanisms of living organisms can use a wide range of fuels—a range that is further increased by the digestive processes which precede respiration.

The release of energy

Compared with the simple combustion that is involved in the working of an engine, the release of energy within the living system is very complex. The heat required to promote rapid oxidation, as well as that actually released during the burning, is far too great for the delicate tissues of plants and animals. So we find that in these organisms energy is released from substrates by a series of gradual stages, not necessarily oxidation, and over a longer period of time.

The chemicals within the body that allow these stages to occur are called *enzymes* (loosely defined as organic catalysts) and they have the effect of lowering the energy required for substances to react together. These enzymes allow the various chemical reactions of the organism to proceed at a high rate yet at low temperatures compared with ignition heats. This can be understood by reference to Fig. 1. Enzymes work most efficiently at approximately 40° C, somewhat above our own body temperature.

The downgrading of the substrates in the body involves many stages, during some of which chemical energy is transferred from the substrate to special energy carriers peculiar to living systems. A typical and very important carrier is *adenosine triphosphate* (ATP) and this occurs throughout living organisms so that we may assume that it became incorporated into the energy exchange system at a very early stage in evolution. We will discuss the role of this substance in the downgrading of glucose in chapter 11, but, in brief, ATP stores the energy in the cells until it is required for muscular contraction, for secretion and absorption or for the manufacture of new substances.

Meanwhile the molecules of the fuel become reduced in size and energy content until, after the downgrading is complete, their remains are rejected from the body as waste products.

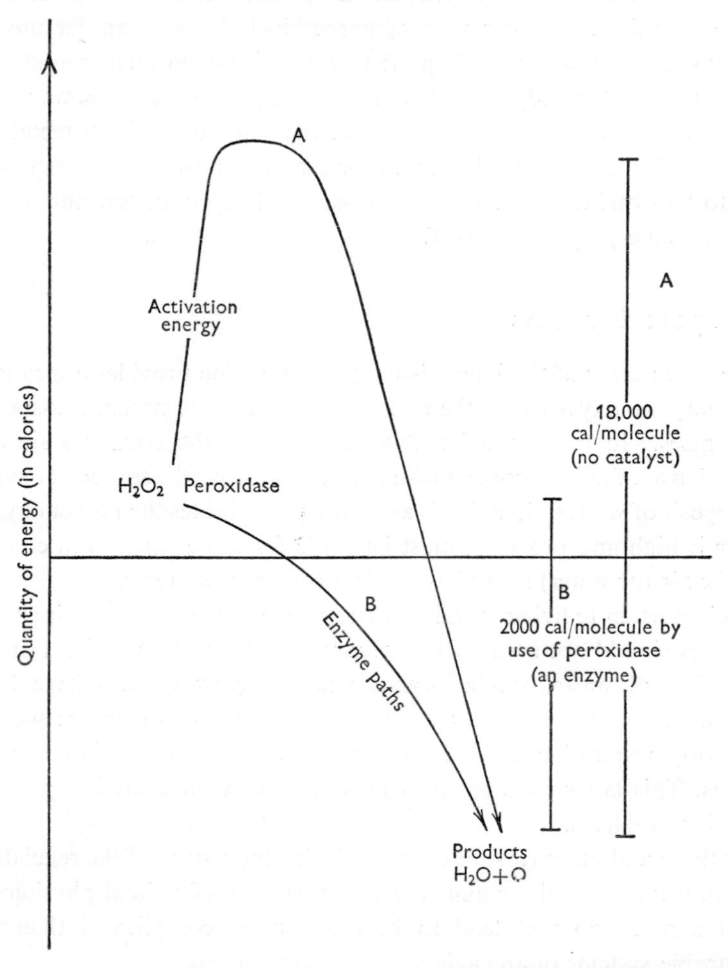

Fig. 1. The dissociation of $H_2O_2 \rightarrow$ water + oxygen with and without the use of the enzyme peroxidase. 1 Cal. = 4·2 kJoules used in SI units.

Energy into work

In the engine, combustion produces heat and movement and the latter is the object of the machine. In animals, although the heat liberated by respiration has been utilised in some cases (as, for example, in birds and mammals where it is an essential factor in their efficient working), the main way in which energy is converted and used is chemical.

One of these conversions of chemical energy results in the combination and shortening of muscle proteins and thus mechanical work can be done by the organism. Further uses of energy include the manufacture (or synthesis) of large molecules, in particular proteins from smaller molecules assimilated into the body as food. The living machine uses all the work that it derives from its respiration for its activity and survival, all metabolic activities being directed to that end. For example, movement allows more food to be obtained, this leads to growth and reproduction and so the living machine perpetuates itself.

Transport Systems

The carburettor of the internal-combustion engine provides a means of combining the oxygen with the fuel while the exhaust system leads away waste gases. In many-celled animals and plants the countless sites of respiration are far from contact with the environment, the source of oxygen and deposit of wastes. In animals as large as vertebrates the rate of oxygen uptake is high and provision must be made for large surfaces in contact with the air (or water) as well as efficient transport systems.

It is important to living machines that the waste products of their energy exchanges should not accumulate because they tend to be toxic. For the elimination of waste, special organs and transport systems have been evolved. Compared with an engine the body produces complex wastes, some from respiration and others from breakdown and replacement of proteins. This latter class contains nitrogen and their excretion presents particular problems.

For these and other reasons connected with the nature of the regulation of a constant internal medium these two aspects of animal physiology, excretion and transport tend to be much more complicated than the comparable systems of an engine.

Coordination

The final function we need to consider is that of the coordination of activities. These activities, as already stated, are directed towards survival and continued functioning of the animal. In mammals, and to a lesser extent in lower animals, the nervous system and hormones provide a means of interpreting the demands made by the external and internal environments. These two systems of coordination bring about adjustments in the body which enable it to adapt itself effectively to the various stimuli that

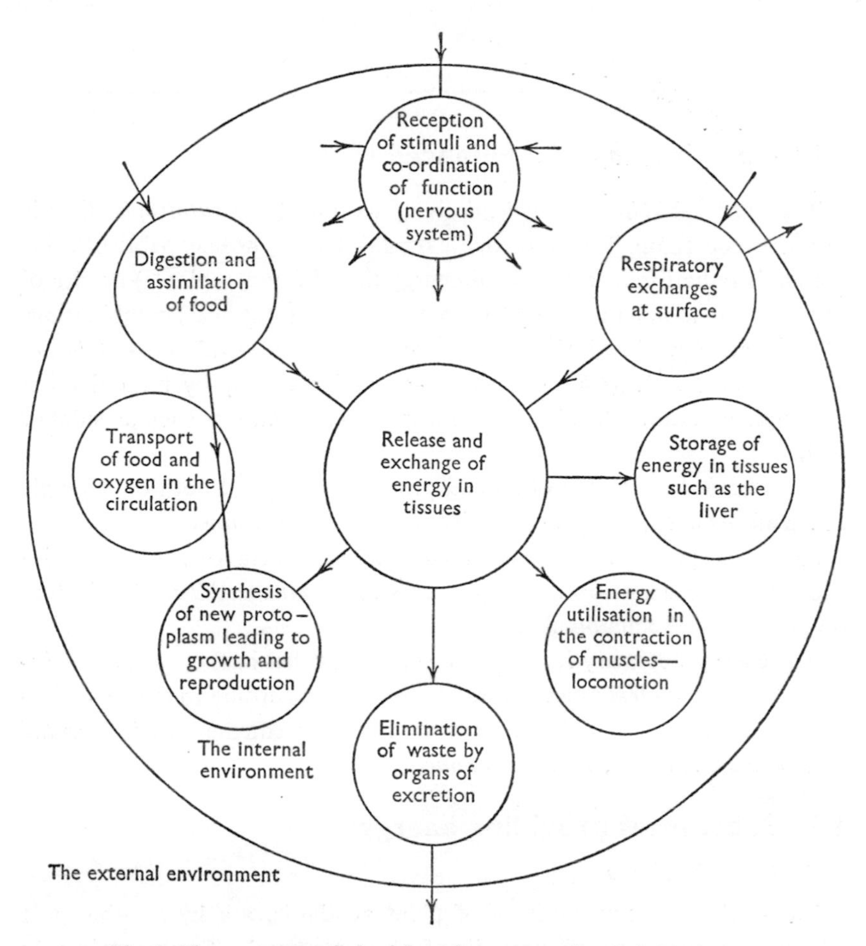

Fig. 2. The processes of physiology in a mammal and their relation to the internal and external environments.

act upon it. Any gross changes are rapidly buffered by feedback mechanisms and all the systems of regulation work towards the restoration of the normal state. As we have seen, this homeostatic concept of regulation and effective adaptation can also be extended to behaviour.

In a non-living machine local controls may be built into the system, but overall functioning remains under human control.

1 NUTRITION

1.1 Constituents of a balanced diet

The mammal is *heterotrophic*, deriving its food from *autotrophs* directly or via a food chain. The autotrophic organisms are those capable of using the sun's radiation as a source of energy (as with green plants) or else of utilising *exothermic* (a chemical reaction that gives out heat) combinations of oxygen with chemicals already present in the earth's crust (*chemosynthetic bacteria*). (*Chemosynthetic*—obtaining energy by the oxidation of simple inorganic chemicals present in the environment such as sulphur or iron.)

The *metabolic activities* of animal protoplasm depend both on a supply of chemical energy and also on an intake of structural substances necessary for enzyme synthesis. These essential building substances, together with fuel and water, make up the constituents of the *balanced diet*, that is, that diet on which the mammal can thrive.

The energy necessary for metabolism is provided in the form of *carbohydrate*, *fat* or *protein* or a mixture of the three, according to the specialisations of the particular animal, while the other ingredients include *essential amino acids*, *vitamins*, *water* and *mineral salts*.

1.2 Substances providing energy

1.21 *Carbohydrates*

These, the primary products of plant synthesis, are by far the most widespread form of energy-providing fuel for mammals. They contain only the three elements carbon, hydrogen and oxygen and are characterised by the ratio of the hydrogen to the oxygen in the molecule being 2:1. As compared with the fats there is also a good deal more oxygen to each atom of carbon. The general structural formula for carbohydrates is $C_x(H_2O)_y$.

Carbohydrates can be classified according to the number of basic sugar or *saccharide* units incorporated in the molecule. This saccharide unit is taken as a *hexose* or *pentose*, that is, a six-carbon or five-carbon atom sugar. However, triose, or three-carbon atom sugars, also exist and may be important intermediates in carbohydrate metabolism. A useful scheme of carbohydrate classification is shown in Table 1 (glycogen is not connected to a disaccharide as it goes directly to glucose on phosphorylation).

Foods with a high carbohydrate content, for example, cereals, rice and potatoes make up the bulk of human diet. Adapted to this diet man has a collection of *carbohydrases* (or carbohydrate-splitting enzymes) capable of dealing with starch and the *disaccharides* maltose, sucrose and lactose. All these are *hydrolysed* in the alimentary canal to glucose or other *monosaccharides* which can be assimilated into the blood. Mammalian herbivores have *symbiotic* organisms in their gut for *hydrolysing* (the splitting of an organic molecule with the insertion of water) the polysaccharides such as cellulose with which they are unable to deal. (Cellulose in the walls of plant cells makes up the bulk of the herbivores diet.)

Table 1

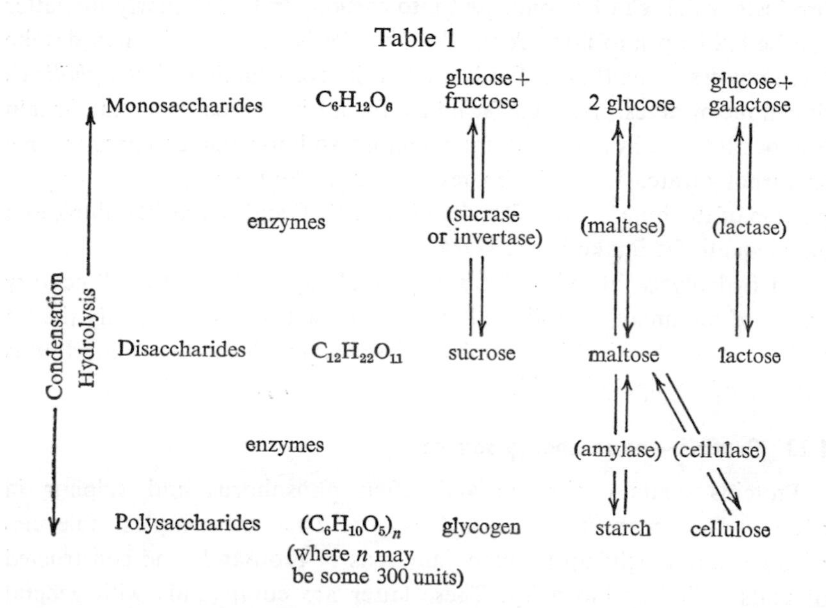

Besides hydrolysing carbohydrates, mammals are also able to synthesise them by condensation reactions involving energy-rich phosphates. Many of the individual carbohydrates are interconvertible and fats may also be built up from monosaccharides.

As distinct from the use of hexoses for fuel the pentoses are incorporated into the structure of nucleoproteins.

On oxidation in the body carbohydrate yields 4·2 Cal/g.

1.22 *Fats*

These contain only carbon, hydrogen and oxygen but unlike carbohydrates there is only a small amount of oxygen in the molecule. They are esters of glycerol and fatty acids. The three fats that are important in human diet are tripalmitin (palmitic acid), tristearin (stearic acid) and triolein (oleic acid). Their empirical formulae are $C_{51}H_{98}O_6$, $C_{57}H_{110}O_6$, and $C_{57}H_{104}O_6$, respectively.

Weight for weight there is more energy content in fat than in other foods. Thus fat provides an excellent fuel store for the body both by its energy content and by the fact that it is inert and can be packed under the skin and round the viscera without interfering with their function.

On metabolism in the mammal fats may yield intermediate CHO products which can be condensed into carbohydrates, similarly the latter can be built up into fats. A small fat intake is necessary in man despite these powers of synthesis. Fat is used in the construction of *phospholipids* of cell membranes. Normal fat intake for an individual in Great Britain is about 100 g daily and fat is not, therefore, so important an energy source as carbohydrate. In cold climates fat may make up a much higher proportion of the human diet. The vitamins A, D, E and K are fat-soluble and an adequate fat intake is necessary.

On hydrolysis fats yield fatty acids and glycerol. In the alimentary canals of mammals complicated means are used to disperse finely the fat which can be absorbed in the unhydrolysed state. A smaller proportion is hydrolysed by lipolytic acid. On respiration fats yield 9·5 Cal/g.

1.23 *Protein—as an energy source*

Proteins contain nitrogen and often phosphorus and sulphur in addition to carbon, hydrogen and oxygen. They are complex molecules of molecular weight up to many hundreds of thousands and constructed of units called amino acids. These latter are compounds with general formula $R.CH_2NH_2COOH$ which become condensed together by peptide links —CO.NH— to form the larger proteins or intermediate substances. The proteins of the mammalian system are in fact composed of various combinations of some 21 amino acids. Some of these 21 amino acids can be synthesised by the body from others, or built up from simpler substances, but in man 10 of them have to be taken in with the diet. For man, therefore, these 10 are called the *essential amino acids* and proteins rich in them are called 1st-class protein. The actual role of these amino acids will be described when we consider proteins as growth and replacement substances.

In the normal diet the bulk of the protein may be 2nd class, that is, not rich in the essential amino acids, and a great deal more protein may be taken in than is required for synthesis of protoplasm. This 'over-consumption' of protein food is true of carnivores but also of Western man with his high standard of living. The excess protein is *deaminated*, that is, its nitrogen content removed, and will also lose sulphur and phosphorus if they are present. These surplus radicals are eliminated as mineral salts or organic excretory compounds. The substance left after deamination still has most of the original energy content, and it will now be a substrate which can be oxidised by carbohydrate pathways.

Thus excess of protein can be utilised as an energy source or as a source of carbohydrate or fat. On the other hand, proteins cannot be synthesised from these latter, and a protein-poor diet represents a real deficiency to mammals.

On oxidation protein yields 5·6 Cal/g.

1.3 The quantity of fuel required

The above classes of foods—carbohydrates, fats and proteins—are all sources of energy to mammals. The amount of energy needed by any one mammal over a particular period will be (*a*) that required to maintain its *basal metabolic activities*, the warmth of its body, its respiratory movements, nervous and secretory functions, etc., and (*b*) that needed to power the actual work done over the same period. Thus, under normal conditions, the mammal is not running, as it were, on capital of energy—in which case it would respire its own stored foods and tissues—but is acting like an internal-combustion engine into which fuel is fed, energy released and work done.

The units in which energy is measured are *Calories* (1 Calorie = 1000 calories, one calorie being the amount of heat required to raise 1 g water 1° C). For a man of average size the basal metabolic requirement is approximately 1780 Cal, a sedentary occupation increases this to some 2420 Cal, while heavy muscular activity can further increase it to over 5000 Cal.

An innate factor determining the basal metabolic rate is the surface area of the body. A naked man loses some 40 Cal/m^2 of body surface per hour. The formula for calculating the body surface area is given by

$$S = 0{\cdot}007184 \times W^{0{\cdot}425} \times H^{0{\cdot}725},$$

where S = surface in m^2, W = weight in kg, H = height in cm.

As far as other mammals are concerned the primary factor determining their energy requirement is their size. Where the animal is small, as for example a mouse, it has a very large surface area in relation to its volume. (This is because surface increases as the square whereas volume by the cubic measure.) Such small mammals have to eat almost continuously to satisfy their energy requirements. Conversely, large mammals such as horses have proportionately a much smaller surface area and therefore heat loss. In both types of warm-blooded animals, mammals and birds, the nature of the exoskeleton, fur or feathers respectively, minimises heat loss. The energy requirements, as shown by oxygen uptake, of various mammals are as follows:

Mouse	2500 c.c./kg/hr
Dog	830 c.c./kg/hr
Man	330 c.c./kg/hr
Horse	250 c.c./kg/hr

1.4 Other constituents of diet

Besides the virtually interchangeable CHO compounds and deaminated proteins described, the mammal requires a small quantity of other substances to maintain its health. The nature of these varies from one mammal to another but in general they include vitamins, essential amino acids, mineral salts and water.

1.41 *Essential amino acids*

These are the amino acids the body must take in for synthesis and growth. They are used directly as structural units, being incorporated with other, non-essential, amino acids into the synthesis of protoplasm. Figure 3 shows how a molecule of pig insulin is made up of essential and non-essential amino acids. In man the essential amino acids are as follows: threonine, valine, leucine, isoleucine, phenylalanine, methionine, tryptophan, lysine and probably histidine and arginine. The non-essential ones include: glycine, alanine, serine, tyrosine, cystine, hydroxyproline, aspartic acid, glutamic acid and citrulline. On the whole animal protein is 1st class,

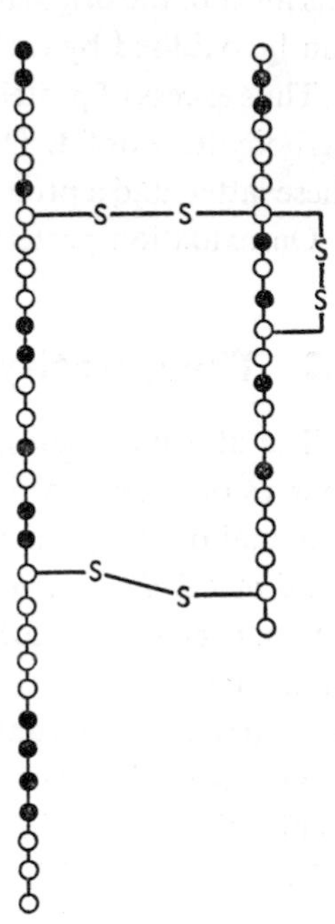

Fig. 3. Insulin molecule, showing incorporated amino acids. ●: Essential amino acid; ○: non-essential amino acid.

and rich in the essential amino acids, while vegetable protein tends to be 2nd class and deficient in these substances.

Probably in man, and certainly in *ruminant* (an animal that can return its swallowed food from a rumen or pouch to its buccal cavity for further mastication by the teeth) mammals, the *intestinal flora* are able to synthesise both essential amino acids and vitamins. These powers may be of considerable importance in the nutrition of the animal; for example, the nitrogen requirement of a sheep can be satisfied by supplying it with urea $CO(NH_2)_2$. It is presumed that the micro-organisms in the gut are able to build up the necessary compounds from this simple source.

1.42 *Vitamins*

These are substances of which small quantities are required to maintain the health of the organism. Thought originally to be amines, it is now known that a number of other types of organic substance are included as vitamins.

The role of vitamins in the cell is different from that of essential amino acids for, whereas the latter are concerned in synthesis, the vitamins act as *coenzymes* directly facilitating metabolic processes. Vitamins are incorporated into the cell's machinery and there bring about some step in respiration, growth and synthesis, but do not themselves become built up into protoplasmic substances.

Vitamins are named after letters of the alphabet but it has been necessary to subdivide the groups and add further ones as new substances are discovered which operate as vitamins. As with the essential amino acids described, the abilities of individual mammal species to synthesise vitamins and hence their requirements vary.

1.421 *Vitamin A*. This is related to carotene (one of the photosynthetic pigments of plants), but this is not so well assimilated from the gut as the pure vitamin. Rich sources of carotene are certain green vegetables, tomatoes, carrots, etc. The vitamin itself is found in animal fats and oils, especially those of fish livers, also in egg yolk, milk and butter fat. In mammals it can be stored in the liver. Some 1·7 mg/day are required and this amount may be doubled in the case of a feeding mother.

Vitamin A is a fat-soluble unsaturated alcohol which is not affected by boiling. It has a number of functions in the body. In the first place the vitamin is a precursor to visual purple in the eye, and a severe deficiency leads to an increase in the threshold of the rods of the retina, that is, more light is required to stimulate them. Secondly, the vitamin is concerned with the laying down of healthy membranes, and where it is lacking, malformed and *keratinised epithelia* (outer cells of the body that have been largely replaced by the protein keratin) result. (Excess of the vitamin causes similar effects.) Where this applies specifically to the cornea of the eye it gives rise to a disease called xerophthalmia, and where to other membranes in the nose and throat, it lowers the resistance of the body to infection. For this reason an increased vitamin A intake is recommended at the start of winter.

Finally this vitamin is concerned, with others, in bringing about normal growth and it is this feature which is used for assay of the vitamin as applied to young rats.

1.422 *The Vitamin B complex.* There are already 12 vitamins of the B complex described. Of these some are of outstanding importance for man and other mammals while others are of more importance for plant growth (e.g. *biotin*).

Thiamine or vitamin B_1 is found in plant seeds, in particular the germ, in yeast and in certain green vegetables. It has a complex structure (formula $C_{12}H_{18}O_2N_4S$) and is not readily affected by boiling. A deficiency of the vitamin causes beri-beri in man and a similar disease called polyneuritis in birds. The diseases are caused by an accumulation of *pyruvic acid* in the cells of the body and by interference in nervous conduction. The muscles become weakened and atrophied though the tissues themselves may become distended with fluid. Thiamine is concerned in carbohydrate respiration and, combined with phosphate, it acts on pyruvic acid formed from *anaerobic glycolysis* (that part of the breakdown of glucose in respiration that requires no oxygen) in the tissues. Normally the pyruvic acid is passed on to an oxidative cycle but this does not take place in the absence of thiamine.

Vitamin B_1 also acts as a growth factor.

Nicotinic acid or vitamin B_2 is found in milk, yeast and eggs. It is necessary to man but can be synthesised by other mammals, such as dogs, provided they are supplied with tryptophan. It is a water-soluble vitamin stable to boiling. The deficiency disease caused by absence of the vitamin is pellagra. Nicotinic acid is concerned in the formation of coenzymes that are important in the activity of *dehydrogenase* (an enzyme that can store or transfer hydrogen from one substance to another) enzymes responsible for removal or transfer of hydrogen during respiration. Again the disease is caused by accumulation of intermediate metabolic products of respiration. Pellagra affects the epithelia and nervous system.

Other effects of B vitamins include partial control of protein synthesis, formation of acetylcholine, correct balance and formation of blood corpuscles and other subsidiary functions.

1.423 *Vitamin C.* This is ascorbic acid, an acid hexose sugar, which is found primarily in citrus fruits and fresh vegetables. The vitamin is water-soluble and easily destroyed on heating.

A deficiency of vitamin C causes scurvy, which is characterised by haemorrhages; that is, an escaping of blood into the tissues. It is thought that the vitamin is concerned in the synthesis of the *mucopolysaccharides* that cement individual cells together. (*Mucopolysaccharides*—organic class of compound that has the general structure of cellulose or starch, i.e. made up of individual glucose molecules, but also contains nitrogen.) Lack of this cement causes the cells to fall apart and the symptoms described. Vitamin C is also involved in the building of connective tissue important in wound healing, in antibody formation and for the health of the teeth.

1.424 *Vitamin D.* Includes a number of fat-soluble substances related to the sterols and *anti-rachitic* (against rickets) in action. In man the vitamin is synthesised by the action of ultraviolet light on the skin and otherwise found in such sources as egg and liver.

Vitamin D provides a means for the assimilation of Ca^{++} and PO_4''' ions from the gut and possibly into the cells of the body as well. Without the vitamin the structures requiring these ions are weakened, this particularly applying to the bones and teeth. Such a condition is known as rickets and is particularly prevalent in temperate climates and slum conditions where sunlight cannot penetrate.

1.425 *Other vitamins.* A number of other vitamins have been described, some for particular invertebrates, some for mammals. Among those important to man are vitamin K which, by regulating the amount of prothrombin in the blood, affects its clotting, and vitamin E, concerned with reproductive processes. Lack of this vitamin may result in habitual abortion of the foetus.

Certain chemicals, related to the vitamins in structure, act as anti-vitamins, replacing the vitamin and disorganising the enzymic processes in which they are involved. These anti-vitamins may provide powerful antibiotics (see page 119).

In general, and as with the essential amino acids, it is supposed that during evolution the mammals, and, of course, other animals, have lost many of the powers of synthesis of ancestral organisms. These substances, the essential amino acids and vitamins, are indispensable in the carrying out of living processes and must be taken in with the diet. This intake may be greatly supplemented by the activity of the micro-organisms of the alimentary tract.

1.43 *Mineral salts*

A large number of different minerals are required to fulfil structural and other functions in the body. These minerals may either be used as ions, as is the case with Na and K, or they may be built up into complex organic substances as with I and Co.

A list of important minerals and some of their uses in the body are:

Ca	bone, teeth, blood
Cl	stomach, blood
P	nerve, muscle, bone
Fe	haemoglobin (incorporated into molecule), cytochrome
Cu	haemoglobin (active as precursor of haemoglobin)
S	hair, nails
F	teeth
Co	bone, blood

Many salts are present as ions in cytoplasm and extracellular fluids and their presence is essential for cell function, also for operation of nervous and muscular tissues. Mineral salts are taken in with the diet, a deficiency of a certain salt sometimes leading to a craving for a food rich in the mineral. Under certain conditions it is possible to suffer from a deficiency disease due to lack of specific salts. Two examples of such diseases are the simple goitre resulting from lack of iodine, and dental caries due to lack of traces of fluorine in the water. However, it should be noted that a disease such as rickets, which involves shortage of Ca^{++} and PO_4''' in the body is due to a vitamin deficiency and not to mineral shortage.

Though each mineral assimilated from the food may have a specific role in metabolism it is the concentration of salts that largely determines the *osmotic pressure* of the body fluids. For normal functioning of the body tissues the osmotic pressure must not vary more than the equivalent of $\frac{1}{10}$ g NaCl in 100 c.c. water. Body mechanisms dealing with correction of osmotic pressure are sensitive to much smaller changes than this. *Ionic regu-*

lation, depending on both intake via the gut and excretion through the kidneys, ensures, as far as possible, that the salt concentration remains constantly within these limits.

Where excessive loss of salt takes place, as in heavy manual work, it is found beneficial to drink dilute NaCl solution. Certain salts, such as magnesium sulphate, which are very slowly assimilated, tend to hold back water in the gut and, by stimulating peristalsis, act as laxatives.

1.44 *Water*

Over and above the fuel and structural items of diet there remains the intake of water to complete the body's requirements. This substance, though partially produced by metabolic processes, must be regarded as a food, as without it the cell, and thus the whole organism, is soon unable to function.

Protoplasm is some 70–80 % water and all its reactions take place in solution. Water is the universal solvent in which the other constituents of living matter can remain either suspended or in solution. The only organisms which contain less than 70 % water are dormant seeds and spores of bacteria and other organisms. In these the reactions within the dehydrated protoplasm are almost at a standstill.

Mammals are some 80 % water and to keep the degree of hydration necessary for the continuance of life they must make up the water lost by evaporation and excretion. A loss of more than 10 % body water is serious, and loss of more than 20 % leads to a thickening of the blood and stoppage of the circulation. The quantity of water passing through the body each day is some 3·5 litres though this may be doubled where excessive sweating takes place. The way these 3·5 litres are made up depends largely on the environmental and other factors, but, on average, some 500 c.c. water per day are lost via the lungs, some 1·5 litres via the kidney and the remaining 2 litres by evaporation from the skin.

Although the amount of sweat and urine lost may vary greatly the latter may not become too concentrated as the nitrogenous waste products it contains are toxic in high concentration.

Of the quantity of water needed daily some half may be taken in as 'hidden' water in the food. Nevertheless, nearly all foods, except soft fruits, require additional water to balance the heat produced in their combustion or excretion.

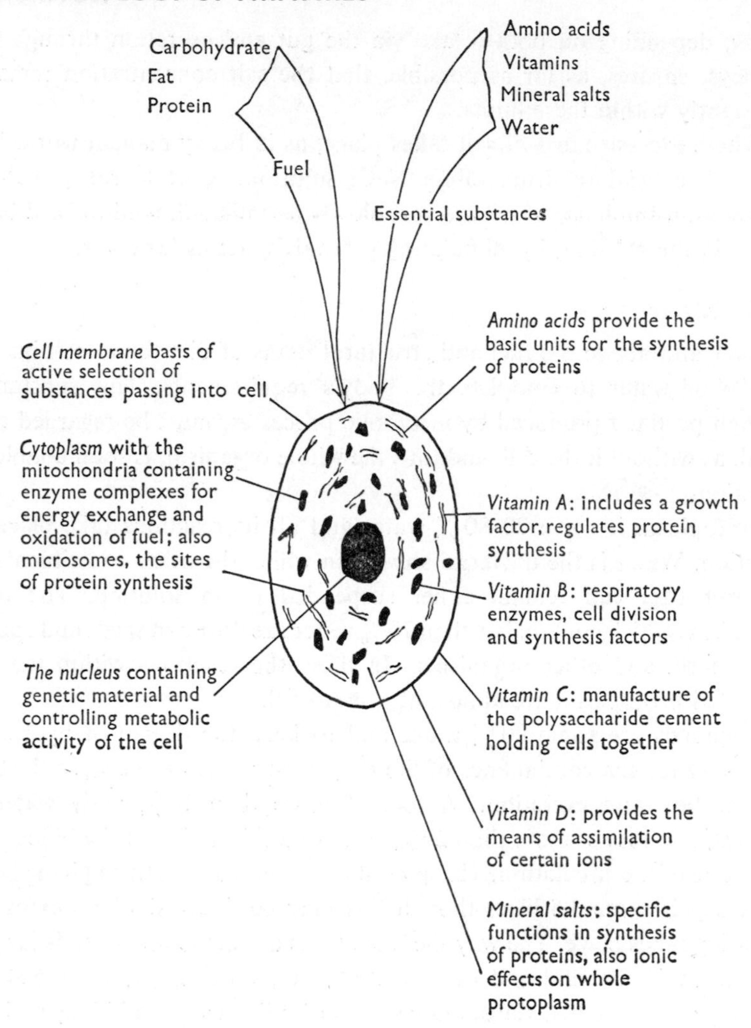
Carbohydrate
Fat
Protein
Fuel
Amino acids
Vitamins
Mineral salts
Water
Essential substances
Cell membrane basis of active selection of substances passing into cell
Cytoplasm with the mitochondria containing enzyme complexes for energy exchange and oxidation of fuel; also microsomes, the sites of protein synthesis
The nucleus containing genetic material and controlling metabolic activity of the cell
Amino acids provide the basic units for the synthesis of proteins
Vitamin A: includes a growth factor, regulates protein synthesis
Vitamin B: respiratory enzymes, cell division and synthesis factors
Vitamin C: manufacture of the polysaccharide cement holding cells together
Vitamin D: provides the means of assimilation of certain ions
Mineral salts: specific functions in synthesis of proteins, also ionic effects on whole protoplasm

Fig. 4. A simplified scheme to show the cell in relation to its nutrition.

1.5 Summary on nutrition

The mammalian body is a living machine which requires the provision of fuel and also of the structural units which maintain and build up the machine itself. These two types of item comprise the balanced diet necessary for health.

The detail of a particular mammal's food will depend on the nature of its specific adaptations, that is, whether it is *carnivorous*, *herbivorous* or *omnivorous*. In their alimentary canals and feeding habits the mammals show great variation and all but a few are fully specialised for the obtaining and utilising of some major food supply—meat, grass, fruit, insects, etc. These adaptations, in so far as they concern the teeth and gut, will be discussed in the next chapter.

Among the various types of mammals a good deal of difference is found in dietary requirements. On the whole carnivores, living on animal protoplasm, are well supplied with the ingredients for their own synthesis, while herbivores depend on the micro-organisms which inhabit their alimentary canals for a supply of vitamins and amino acids. The synthetic powers of mammals are not very great compared with those of lower organisms and, by one means or another, it is necessary for them to select or provide themselves with the numerous constituents of their balanced diets. The nutrition of these animals illustrates the biological dictum that 'increase of morphological complexity is often accompanied by decreasing powers of biochemical synthesis'.

DIGESTION

2.1 The nature of digestion

Digestion is the means whereby the various items of the diet become broken up into a form in which they can be assimilated into the blood or lymph. The breaking down of large molecules that takes place in digestion is based on the chemical reaction of hydrolysis, whereby water is inserted across the junctions of the initial molecule making many smaller units. Typical examples of hydrolysis, as it occurs in digestion, are shown in Table 2.

Table 2

Carbohydrate

$$\underset{\text{Starch}}{(C_6H_{10}O_5)_n} + \frac{n}{2}H_2O \xrightarrow{\text{Amylase}} \frac{n}{2}\underset{\text{Maltose}}{C_{12}H_{22}O_{11}}$$

Fat

$$\begin{array}{l} C_{17}H_{35}COO.CH_2 \\ \qquad\qquad\quad | \\ C_{17}H_{35}COO.CH + 3H_2O \xrightarrow{\text{Lipase}} 3\underset{\text{Stearic acid}}{C_{17}H_{35}COOH} + \underset{\text{Glycerol}}{CH_2OH.CHOH.CH_2OH} \\ \qquad\qquad\quad | \\ C_{17}H_{35}COO.CH_2 \\ \qquad \text{Tristerin} \end{array}$$

Protein

$$R.CH_2CO.NH.CH_2COOH + H_2O \xrightarrow{\text{Protease}} R.CH_2COOH + NH_2CH_2COOH$$

The enzymes that bring about these reactions are capable also of catalysing the reverse *condensation* reactions. In fact, although the enzymes control the rates of reactions, the final concentrations are at least partly determined by the fate of the products. Thus in the case of sucrose hydrolysis in the duodenum the products, glucose and fructose, are taken into the blood vessels of the ileum by a process of active absorption, so that, by the Law of Mass Action,[1] the reaction tends towards the formation of more of these products. The reverse reaction for the condensation of monosaccharides into more stable sugars takes place in the liver and for this, since it is a building-up process, energy must be supplied. (Hydrolysis reactions yield energy.)

Whereas in certain lower animals the food particles may be taken directly into the cells lining the gut (called *intracellular digestion*), in mammals and other vertebrates the enzymes are liberated into the lumen of the gut and the products of their action absorbed (*extracellular digestion*). In digestive systems of the latter type two regions are usually differentiated,

[1] The rate of a chemical reaction is proportional to the concentration of each of the reacting components.

one specialised for digestion and a further one for absorption. Within the first of these regions the various types of enzyme-producing cell have become grouped together so that concentrated solutions acting at an optimum pH can attack the food. In mammals digestion begins in the buccal cavity, where the teeth are differentiated according to the type of food the mammal eats and salivary enzymes start the chemical digestion of food materials.

The alimentary canal can be considered as a tube whose lining and properties are more appropriate to a surface structure than an internal system. The cavity of the gut is quite distinct from the true body cavity or *coelom*, which is surrounded by mesoderm. Within the gut flourishes a vast bacterial flora, some 40 % of human faecal matter is bacteria, and this flora is normally commensal or symbiotic, that is, non-harmful or actually useful, provided it remains in the food canal. On the other hand, entry of micro-organisms into the true body cavity leads to immediate infection and disease. There is no one typical mammalian alimentary canal but on the whole those of herbivores are more complex than others. A mammalian herbivore depends on its internal flora to carry out the bulk of its digestive processes and its guts are specially adapted to house micro-organisms and encourage their activity.

If we consider first the human alimentary canal and then some of the modifications found in other mammals and vertebrates a clear picture of the mechanism and coordination of the digestive system can be built up. The vertebrate alimentary canal is organised on a uniform basis whose main parts are listed on page 22.

2.2 The human alimentary canal

The mouth leads into the buccal cavity in which are the teeth and into which lead ducts from the salivary glands. As is characteristic of mammals the teeth are differentiated in accordance with the diet, which, in the case of man, is a variety of plant and animal food. Human teeth occur as laterally flattened *incisors* at the front, rather small *canines*, the eye teeth, outside the incisors and then *premolars* and *molars* at the back of the jaw. In the case of omnivorous dentition, the differentiation of the various types of teeth is not well marked. Before the adult condition is reached a milk dentition exists and may be summarised by the following dental formula. (The dental formula is an expression for the teeth on one side of the mouth.)

$$I\tfrac{2}{2}C\tfrac{1}{1}.PM.\tfrac{2}{2}M\tfrac{0}{0} = 20.$$

Each tooth develops from an *odontoblast* or tooth bud. This secretes a layer of prismatic calcium carbonate or enamel. Below this, forming the bulk of the tooth, is *dentine*, a substance related to bone and permeated with minute canals in which run nerve endings and blood capillaries originating from the *pulp cavity*. Fixing the tooth into the socket of the jaw is a further calcareous substance, *cement*.

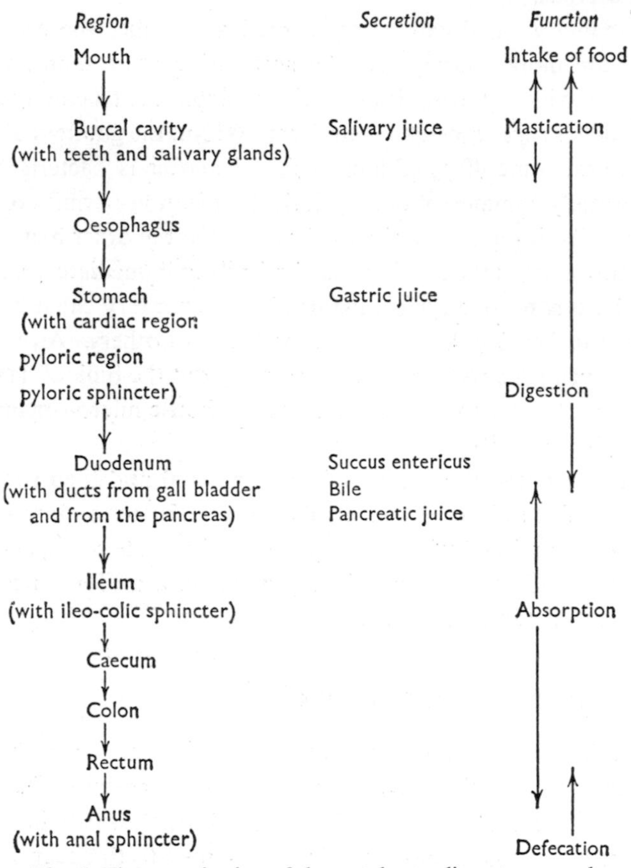

Fig. 5. The organisation of the vertebrate alimentary canal.

Human teeth have small canals leading up through the roots to the pulp cavity. These control the amount of food entering the tooth and thus its rate of growth, which in the case of man is very slow.

Man has 32 teeth, midway between the herbivorous and carnivorous condition. His *dental formula* is as follows:

$$I\tfrac{2}{2}C\tfrac{1}{1}P.M.\tfrac{2}{2}M\tfrac{3}{3} = 32.$$

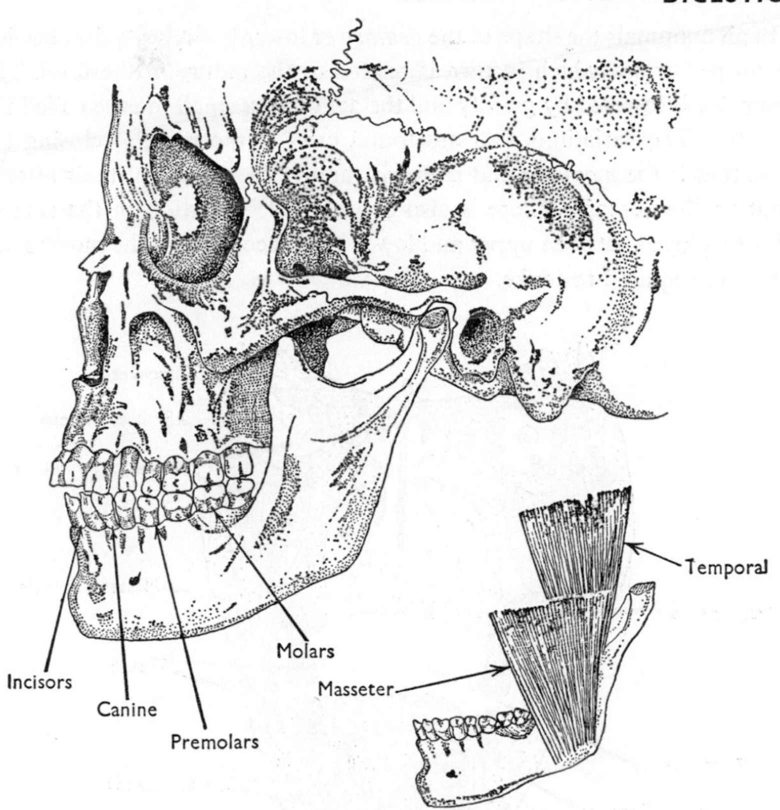

Fig. 6. Human skull showing dentition.

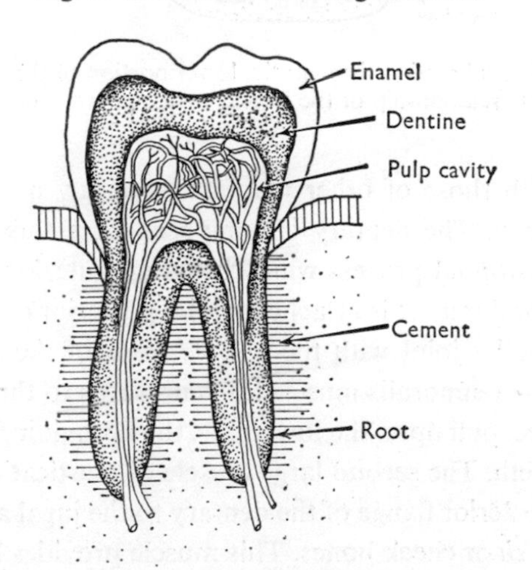

Fig. 7. Longitudinal section of human molar.

In all mammals the shape of the *dentary* or lower jaw reflects the chewing action performed which in turn is related to the nature of the food. The lower jaw is lowered by gravity and the action of a small muscle called the *digastric*. The variations that are found exist in the muscles closing the jaws, that is the *masseter* and *temporal*, and in the nature of their attachment to the dentary. There is also considerable variation in the sort of joint employed between upper and lower jaws according to the movements these are required to make.

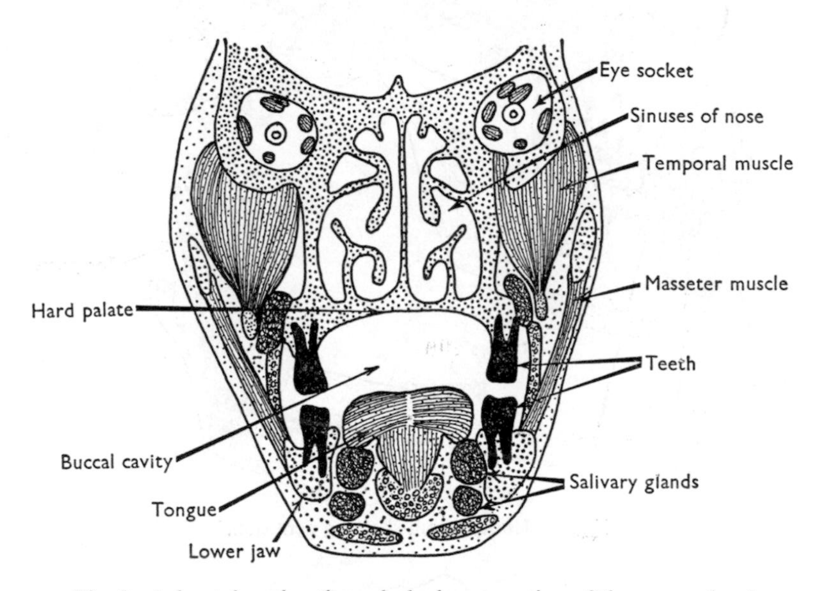

Fig. 8. A frontal section through the lower portion of the numan head showing relationship of the buccal cavity to jaws and glands.

Compared with those of other mammals the jaw muscles of man are somewhat reduced. The dentary is composed of a tooth-bearing ramus and a vertical coronoid process which bears an anterior angular process, where the temporal muscle is attached, and a posterior condyle. This latter makes a loose roller joint with the *glenoid fossa* of the *squamosal* of the skull. In man the temporalis muscle leads upwards to the side of the head and exerts a direct pull upon the lower jaw, thus providing the biting action of closing the teeth. The second large muscle of mastication, the masseter, runs from the posterior flange of the dentary to the jugal and squamosal of the *zygomatic arch* or cheek bones. This muscle provides both vertical and lateral movement to the dentary and powers the chewing action of the molar teeth. Both these muscles are of similar size and importance and work at a similar mechanical advantage in the distance of their points of attachment from the articulation.

In the size of the coronoid, the muscle arrangement, the nature of the articulation and pattern of the teeth, man is intermediate between herbivore and carnivore. These more specialised mammals will be described at the end of the section on the human alimentary canal.

2.3 The principles of coordination of secretion and digestive movements

Both nervous and endocrine mechanisms are used in the coordination of the digestive system. That part of the nervous system concerned in the operation of the viscera is called the *autonomic* and itself consists of two systems, the *sympathetic* and *parasympathetic*. The former is an extension of the segmental motor paths of the visceral spinal arcs and is stimulated by the hormone *adrenaline* (it is dealt with more fully in chapter 8).[1] Its basic function is to increase the efficiency of the body for immediate action, and besides the positive actions it produces (increase in respiration, blood pressure, release of glycogen, etc.) it also tends to inhibit digestive activity. Thus the connections from the sympathetic system which go to the parts of the alimentary canal or its associated glands depress their activity, or, under extreme stimulus, cause a reversal of the normal gut movements of *peristalsis*. Sympathetic stimulation also causes a reflex emptying of the bladder and bowel.

The parasympathetic system consists of fibres in the vagus cranial nerve, Xth, together with fibres of the trigeminal nerve, Vth, and the facial nerve, VIIth, as well as part of the pelvic plexus. The general action of the parasympathetic system is to stimulate digestion and secretion.

Besides these two parts of the nervous system and their associated endocrines a number of localised hormones are used in the coordination of digestion. These usually fortify the secretions of a particular gland under the stimulus of food but may also set a gland in action on the approach of food as with the effects of duodenal secretion on the pancreas. Local nerve reflexes are also used to bring about peristalsis, villi movements and all the activities concerned in the passage of food and elimination of waste through the gut.

[1] This section will be more readily understood after the autonomic nervous system has been studied (p. 210).

The general picture is thus of a complex system continuously breaking down and assimilating the food passing into the mouth of the animal. This system works largely under its own local control measures and yet in time of stress its activity ceases, becoming subordinated to the survival of the whole organism.

2.4 The organisation of the alimentary canal

As has already been stated the gut is a long tube the parts of which have been variously specialised for different functions. In general the upper regions have conditions favourable for the operation of specific enzymes while the lower parts with their enlarged inner surfaces are specialised for assimilation.

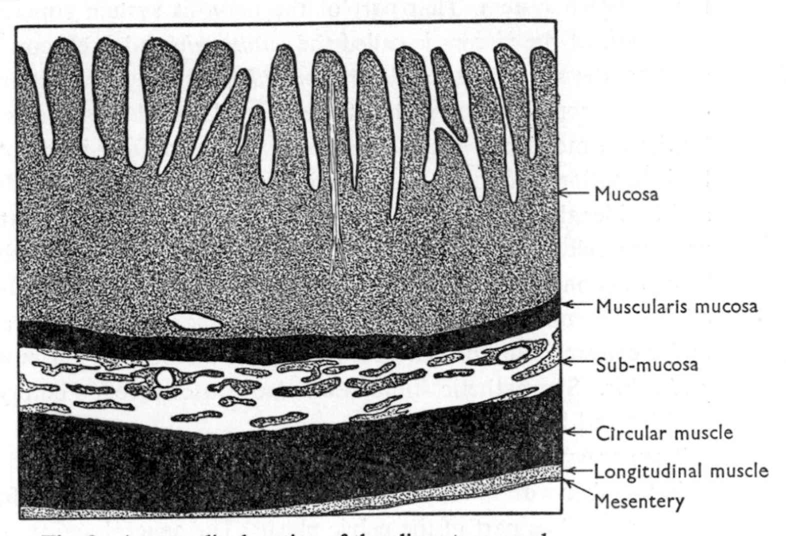

Fig. 9. A generalised section of the alimentary canal.

A generalised transverse section of the alimentary canal is shown in Fig. 9. Once this is clearly understood it becomes much easier to follow its modifications.

2.41 *The buccal cavity*

The region between the mouth and the oesophagus is called the buccal cavity. Around the top it is bounded by the palate, at the base by the tongue, and at the sides by the muscles of the cheeks (buccinators). The cells lining the cavity form a *stratified epithelium* as distinct from the *columnar epithelium* found further down the gut.

Into the buccal cavity empty three pairs of salivary glands, the sublingual, submandibular and parotid, which together secrete the digestive juice, saliva. Posteriorly the soft palate acts as a valve preventing food getting into the back of the nasal cavity, while the entrance of the windpipe, or trachea, is provided with another valve, the epiglottis, which stops food passing into it. Reflex arcs involving the facial, VIIth, glossopharyngeal, IXth, and vagus, Xth, nerves ensure that the *bolus* of food formed by the tongue is passed smoothly down into the oesophagus during swallowing.

The buccal cavity is well provided with sensory apparatus to test the substances passed into the mouth. Both distant and immediate perception of *olfactory* (concerned with smell) stimuli cause the secretion of saliva as does stimulus of the taste organs of the tongue. These stimuli are transmitted as sensory impulses from the olfactory nerve, Ist, the trigeminal, Vth, and facial nerves, VIIth, to the medulla of the hind brain, thence via motor nerves to the salivary glands. The presence of material in the mouth also stimulates secretion and from all these sources of stimulus pathways leave via the vagus, Xth, nerve which leads to the stomach. The latter is thus prepared in advance to receive the food. It should be noted that stimulation of the sympathetic system under conditions of fear or shock causes small quantities of viscid saliva to be produced and inhibits the normal copious flow.

Saliva itself consists of sodium bicarbonate (which makes an alkaline medium in which the salivary *amylase* or *ptyalin* can work), the ptyalin, water and mucus. The ptyalin hydrolyses carbohydrate starch to malt sugar or maltose:

$$(C_6H_{10}O_5)_n + \frac{n}{2} H_2O \rightarrow \frac{n}{2} C_{12}H_{22}O_{11}.$$

This change takes place in a number of steps of hydrolysis yielding soluble starch, *erythrodextrin*, and *achroodextrin*. This reaction can be followed in detail as the original starch gives a purple colour with iodine, while erythrodextrin gives a red colour and the achroodextrin produces no colour. Besides this digestive action of saliva the mucin lubricates the food and assists in swallowing. Meanwhile the teeth and tongue help to mix the enzyme with the food and although carbohydrates are the only substances actually hydrolysed in the buccal cavity the *mastication* does greatly assist the subsequent digestion of all foodstuffs by rendering it into small particles upon which the enzymes can act.

2.42 *The oesophagus*

This is a tube some 10 in. in length connecting the buccal cavity to the stomach. The swallowing reflex relaxes the upper part of the oesophagus and a wave of relaxation followed by one of contraction pushes the food bolus down to the stomach. This sort of movement is common to all regions of the gut and is called peristalsis. It depends, normally, on local nerve reflexes triggered by the presence of food relaxing and contracting the longitudinal and circular muscles alternately, the contraction following the food and the relaxation preceding it. The innervation of the oesophagus, however, is derived from those parts of the nervous system concerned in swallowing and not by local reflex.

2.43 *The stomach*

The stomach is a muscular bag, capable of great distension, which is divided into *cardiac* and *pyloric* parts and leads into the duodenum via the *pyloric sphincter*. The active churning movements of the stomach take place in the lower portion, the food being passed into this by the peristaltic contractions of the upper part. These movements mix the stomach contents with the enzymes, rendering the whole into a semi-liquid mass known as *chyme*.

The walls of the stomach are much folded and covered by a layer of columnar epithelium. Indented into the *mucosa* are some three million *gastric pits* or tubules (Fig. 10). Specialised cells in the tubules secrete the gastric juice whose composition and action is as follows:

(*a*) A copious layer of mucus is secreted by the upper cells of the gastric pits and the lining cells of the stomach. Mucus is a constituent of all intestinal juices and protects the tissues from the action of enzymes. It is amphoteric (a substance that can act as both acid and base according to conditions) in nature and can thus neutralise both acids and alkalis besides providing a mechanical buffer between hard food substances and the living tissues.

(*b*) Hydrochloric acid is secreted to give a final concentration in the stomach of approximately $N/10$ which gives a pH of 2. The effects of this acid are to bring about direct hydrolysis of food substances and to provide a favourable medium for the activity of the enzyme pepsin. The acid also kills the majority of the bacteria taken in with the food. The mechanism whereby the acid-producing, or *oxyntic*, cells of the gastric

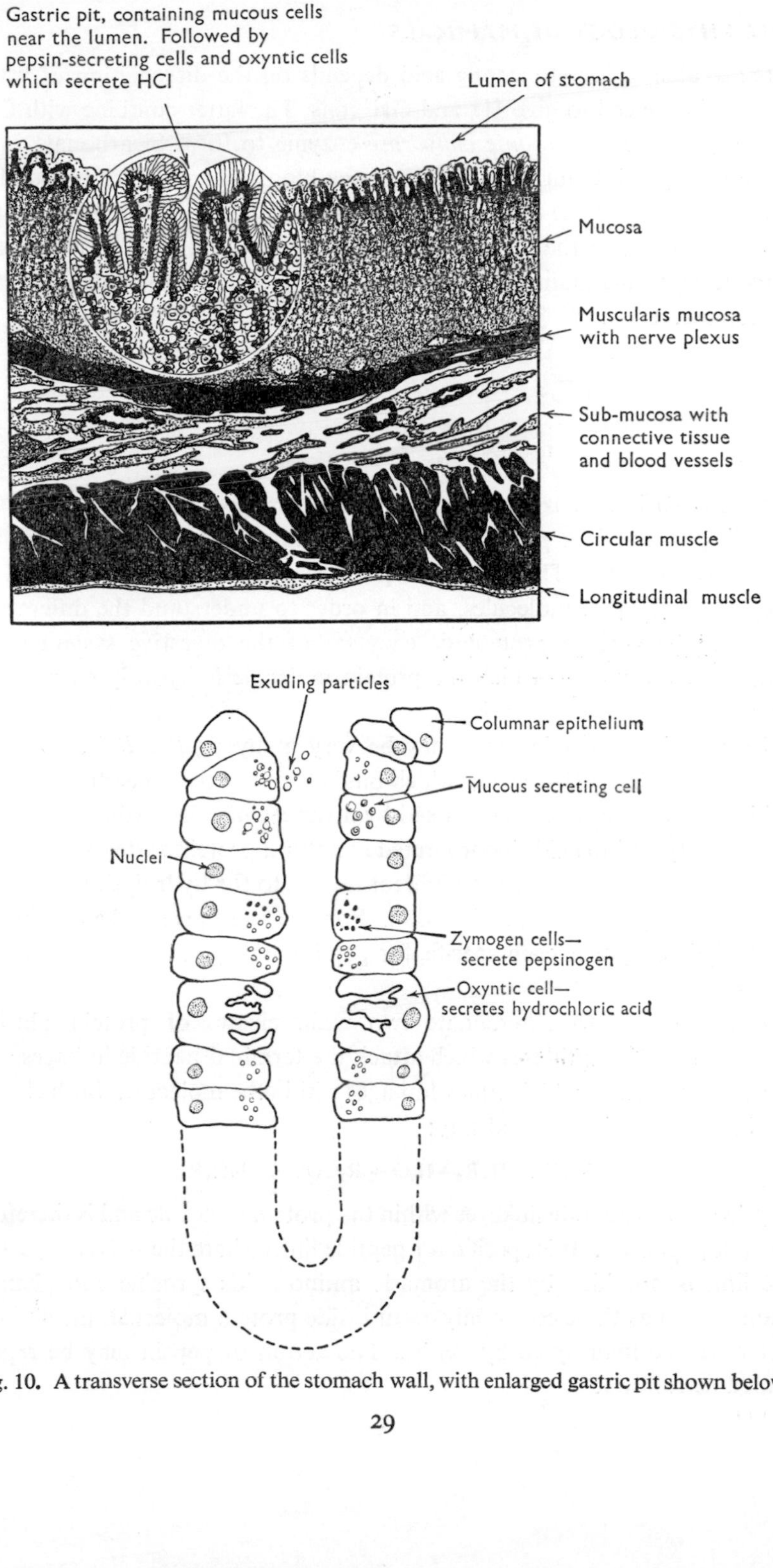

Gastric pit, containing mucous cells near the lumen. Followed by pepsin-secreting cells and oxyntic cells which secrete HCl

Lumen of stomach

Mucosa

Muscularis mucosa with nerve plexus

Sub-mucosa with connective tissue and blood vessels

Circular muscle

Longitudinal muscle

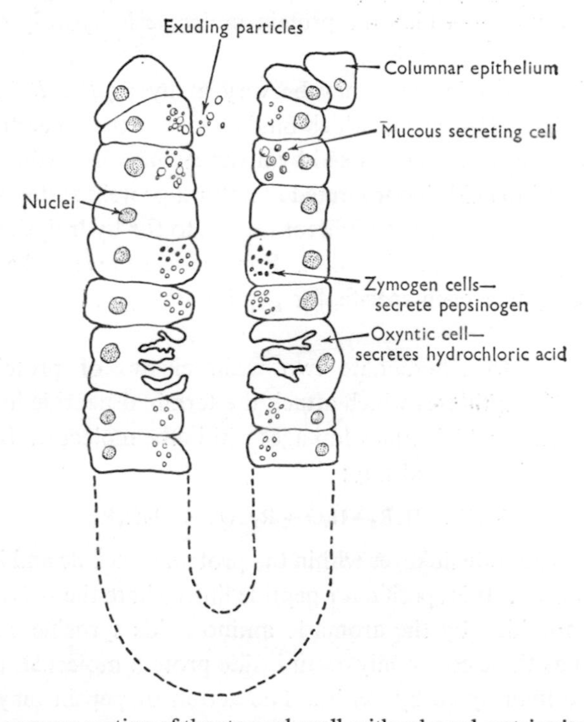

Fig. 10. A transverse section of the stomach wall, with enlarged gastric pit shown below.

pits are able to secrete strong acid depends on the dissociation of water present in the cell to give H^+ and OH′ ions. The latter combine with CO_2 in the presence of *carbonic anhydrase* enzyme to form bicarbonate ions, these then passing out of the cell into the bloodstream, which is found to become more alkaline shortly after the intake of food. Meanwhile the H^+ ions diffuse out of the cell into the lumen of the tubule or pit and at the same time, to maintain the neutrality of charges across its membrane, Cl′ is secreted. Thus:

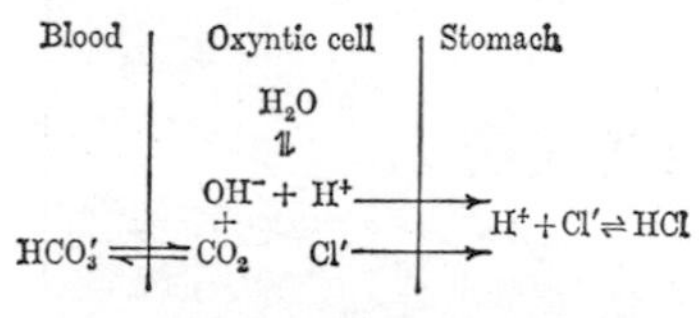

(*c*) *Pepsin* is an enzyme secreted by the zymogen or *peptic cells* of the gastric pits and is a *proteolytic* enzyme acting in the acid medium provided by the HCl. The nature of protein digestion is very complex as they are such large molecules, and in order to understand the differences between the various proteolytic enzymes of the digestive system some outline of the way in which the protein molecule is hydrolysed must be given.

In any given protein there may be very many *peptide linkages*, these being the —CO.NH— group which binds together the constituent amino acids. These groups may be flanked by a variety of others (there are more than twenty amino acids incorporated in natural protein systems) and each peptide linkage may present a different surface to the hydrolysing enzymes, which are very specific in their action. For this reason we find a number of proteolytic enzymes, each specific for peptide linkages, enclosed by other groups.

It is possible to differentiate two main classes of protein-splitting enzyme, the exopeptidases which attack the terminal peptide linkages and the endopeptidases which attack linkages within the molecule. Both classes act in the following general way:

$$R_1CO.NH.R_2 + H_2O \rightarrow R_1COOH + NH_2R_2.$$

Pepsin attacks peptide linkages within the protein molecule and is therefore an endopeptidase. It is specific for peptide links where the —NH— part of the link is provided by the aromatic amino acids tyrosine and phenylalanine, and as these commonly occur inside protein molecules most of the latter will be hydrolysed by pepsin. The action of pepsin may be represented as follows:

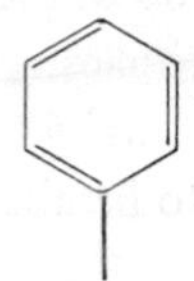

(Further amino acids) R.CO ⁝ NH.CH.CO.NH.R (further amino acids)

Pepsin cleavage here

Hydrolysis by pepsin leaves the protein as *peptone* molecules. These will contain a number of amino acids which will be less than the protein from which they are derived.

As with other proteolytic enzymes pepsin is secreted as its precursor pepsinogen and is activated by HCl and then *autocatalytically* (it catalyses or brings about its own hydrolysis) to pepsin.

(*d*) *Rennin* is also found in gastric juice, particularly in young mammals, and its function is to convert, in the presence of Ca^{++} salts, the soluble *caseinogen* protein of milk into insoluble *casein*. The solubility of casein depends on pH and a further change, 'clotting', takes place under the action of rennin and Ca^{++}.

Coordination of the stomach takes place initially by impulses passing down the vagus nerve (Xth) into the nerve net at the base of the mucosa. These are caused by the presence of food in the buccal cavity or by the smell or sight of food. Once the food actually reaches the stomach it causes the release of a hormone *gastrin* from the stomach epithelia into the blood. This hormone passes rapidly round the bloodstream and, circulating in the tissues of the gastric pits, increases their activity. The action of the sympathetic connections to the stomach tends to depress enzyme secretion although not necessarily acid secretion.

After some 4–5 hr in the stomach the semi-digested matter, called *chyme*, begins to pass via the pyloric sphincter into the duodenum. The actual opening of the sphincter muscle is possibly due to a change in the acidity or osmotic strength of the stomach contents at the end of digestion. This chemical change relaxes the sphincter so that the peristaltic contraction of the stomach pushes across it and thus passes food into the duodenum.

By the time this happens pepsin and rennin have done their work and some hydrolysis of food substances has taken place in the presence of HCl. The latter has also broken down connective tissue in the food, releasing its digestible contents. The warmth of the stomach has melted any fat present and the continuous churning action of the stomach wall has done much to reduce the food to very small particles.

Substances of small molecular weight such as monosaccharide sugars and ethyl alcohol may also start to be absorbed through the mucosa of the stomach.

2.44 *The duodenum*

This is the next region of the gut and is the first part of the small intestine. The same general tissues can be recognised in the wall of the duodenum as are found in the stomach (Fig. 11). The outer layers consist, as before, of mesentery, of longitudinal and circular muscles followed by

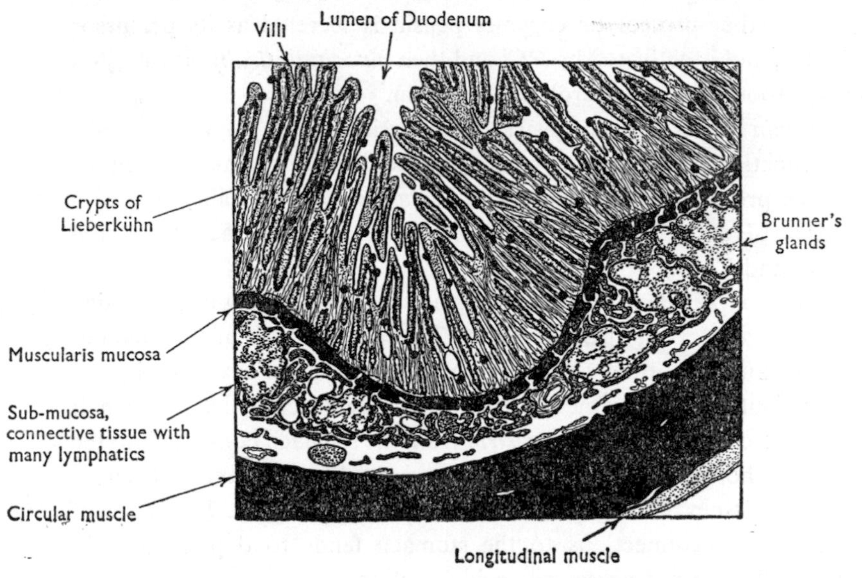

Fig. 11. A transverse section of the duodenum wall.

a nerve plexus and then the sub-mucosa. The latter is full of the characteristic *Brunner's glands* whose presence distinguishes the duodenum from other regions of the small intestine. In the mucosa there are many secretory *crypts of Lieberkühn* while the surface of the mucosa consists of many folds called *villi.* These effectively increase the surface area in contact with the food. Into the duodenum opens the *hepato-pancreatic duct* which brings important digestive juices from the *gall bladder* and the *pancreas.* As these secretions act upon the food at the beginning of the duodenum and before the *succus entericus*, or digestive juices, of the latter, their composition and function will be described first.

2.441 *The bile.* This originates from the liver where it is passed across from the canals of the lobules (Figs. 60 and 61) into the small intracellular *bile canaliculi.* These collect up into larger vessels which drain into the gall bladder, embedded within a lobe of the liver, and from the bladder the bile duct runs down to meet the pancreatic duct. Bile consists of certain salts such as *sodium glycocholate* and *sodium tauroglycocholate* which reduce surface tension and are important in *emulsifying* fats. This process makes them more readily digested by the lipase from the pancreas which will only function normally in the presence of these salts. The bile also contains break-down products from blood haemoglobin which are *bilirubin* (red) and *biliverdin* (green) which colour the faeces. A quantity of HCO_3' is released with the bile fluids giving them a pH of 8 and helping to reduce the acidity of the contents of the duodenum. Bile also stimulates the peristaltic movements of the gut.

2.442 *The pancreas.* A diffuse pink organ found in the first loop of the small intestine after it leaves the stomach. It produces important digestive enzymes, termed collectively *pancreatic juice,* and is also an endocrine organ. These two functions are reflected in the histology of the pancreas where the *islets of Langerhans,* which produce the *insulin,* can be distinguished from the ground mass of secretory tissue where the enzymes are formed. Collecting ducts join up into the pancreatic duct and this, joining with the bile duct, enters the top of the duodenum.

The pancreatic juice contains the following enzymes:

(*a*) *Trypsinogen* is an *endopeptidase* acting on peptide links within the protein molecule. It is activated by *enterokinase* from the duodenum as well as autocatalytically by itself and turned to *trypsin.* In this active form it operates best at pH 8 though it will show less activity in more acid conditions.

Trypsin attacks peptide linkages where the —CO part of the link is provided by the amino-acids arginine or lysine, e.g.

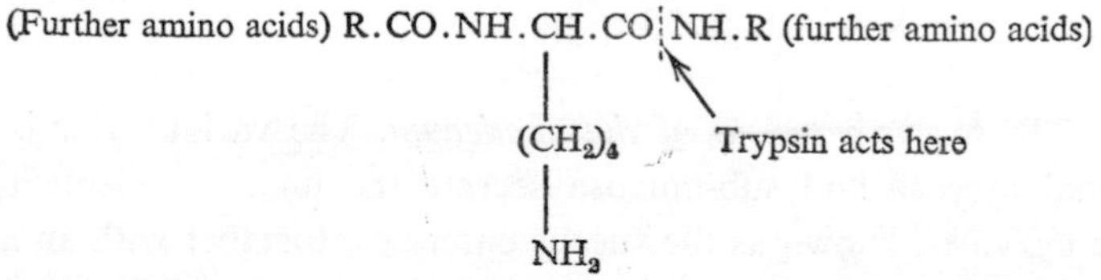

(Further amino acids) R.CO.NH.CH.CO ¦ NH.R (further amino acids)

$(CH_2)_4$ — Trypsin acts here

NH_2

It converts proteins or peptones to di-, tri-, or polypeptides having 2, 3 or more amino acids in their respective molecules.

(*b*) *Chymotrypsinogen* is activated by trypsin and like the latter is an endopeptidase. It cleaves peptide links wherever the amino acids phenylalanine or tyrosine provide the —CO part of the link, and like trypsin results in the hydrolysis of proteins and peptones to smaller units.

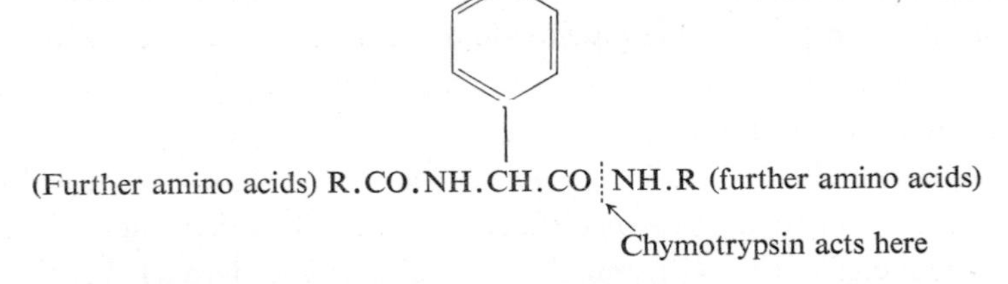

(*c*) A *carboxypeptidase* is an exopeptidase which acts on peptide linkages next to a carboxyl group. It works on small protein units (peptones or polypeptides) after they have been released by endopeptidases (Fig. 12).

(*d*) A *lipase* or a fat-splitting enzyme, hydrolysing the links between glycerol and fatty acids to release these substances in the duodenum. A fairly typical hydrolysis would be:

$$\underset{\text{Tristearin}}{C_{57}H_{110}O_6} + 3H_2O \rightarrow \underset{\text{Stearic acid}}{3C_{17}H_{35}COOH} + \underset{\text{Glycerol}}{CH_2OH.CHOH.CH_2OH}$$

Some of the fats present may not be hydrolysed, however, and will be assimilated directly while much of it may be converted to *glycerides* which contain glycerol and small $—(CH_2)_nCH_3$ side chains.

(*e*) An *amylase* is similar in action to the salivary ptyalin already described and it completes the hydrolysis of starch started by the latter. In the duodenum amylase turns starch to the disaccharide maltose which is later broken down to monosaccharides.

Amylases exist in α and β forms and these attack starch molecules in different ways—but both bring about hydrolysis.

All the enzymes of the pancreas work best at a pH of 8–9 and towards this end alkaline fluids are secreted in the bile and from the pancreas and duodenum (see, however, p. 35 (*f*)).

2.443 *The succus entericus of the duodenum.* The various glands of the duodenal mucosa and sub-mucosa secrete the final complement of digestive enzymes, known as the succus entericus, together with an alkaline fluid and mucus. The enzymes come from the crypts of Lieberkühn at the base of the villi while the alkaline fluid and mucus are mainly derived from the Brunner's glands in the sub-mucosa. These glands are characteristically found in the upper part of the duodenum where the need to neutralise stomach acid is greatest. The succus entericus contains the following:

(*a*) *Enterokinase.* This activates the trypsinogen from the pancreas by hydrolysis of a terminal peptide link turning it to active trypsin. This means that the latter enzyme is only in its active form at the site of its activity and in the presence of food in the duodenum.

(*b*) *Erepsin* is really a number of different enzymes which complete the digestion of the protein molecule. After the activity of trypsin and chymotrypsin described above the proteins are in the form of *polypeptides* which may consist of small numbers of amino acids linked together, say 2–6. Erepsin consists of various enzymes each of which is specific for a particular sort of polypeptide. Thus one that acts on a peptide linkage next to a —COOH is called a *carboxypeptidase,* while another type acting on the peptide linkage next to an NH group is called an *aminopeptidase.* Finally the polypeptides are reduced to groups of only 2 or 3 amino acids and these are finally hydrolysed into single amino acids by dipeptidases and tripeptidases respectively. All these enzymes are included in the term erepsin. A summary of protein digestion is given below.

(*c*) *Lipase* acts on the hydrolysis of fat as already described for pancreatic lipase.

(*d*) *Sucrase* acts on the disaccharide sucrose, or cane sugar, converting it to glucose and fructose. This enzyme is also called *invertase.*

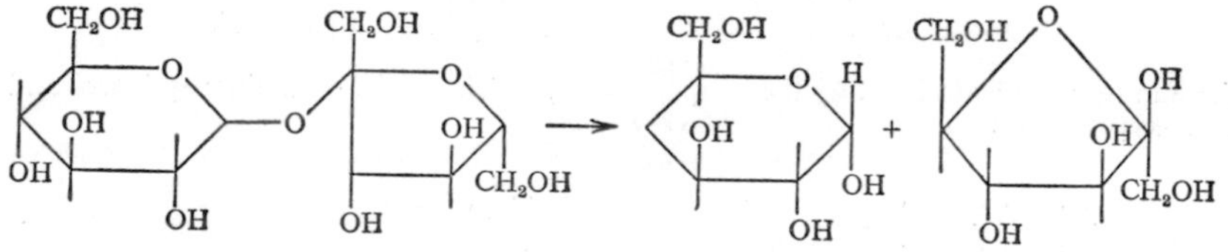

(*e*) *Maltase* acts on the disaccharide maltose, or malt sugar, turning it to two molecules of glucose.

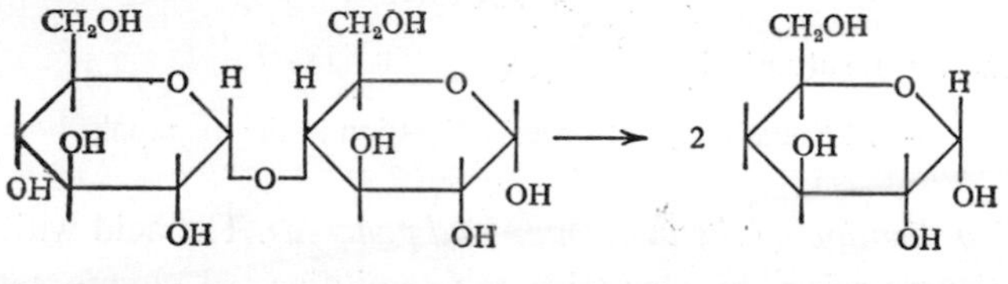

(*f*) *Lactase* converts the disaccharide lactose, or milk sugar, to glucose and galactose.

These duodenal enzymes act best in a medium of pH 8·3, but owing to the very acid state of the stomach the general pH of the duodenum is about 5·5. It may reach 6 in the ileum but alkalinity is never obtained.

All classes of food substance have now been broken down into small molecules which can be assimilated into the bloodstream or lymph. The carbohydrates are in the form of glucose or other monosaccharides, the fats remain as fat, or have been hydrolysed to fatty acids, glycerol or glycerides, and proteins are in the form of amino acids.

A theoretical protein molecule

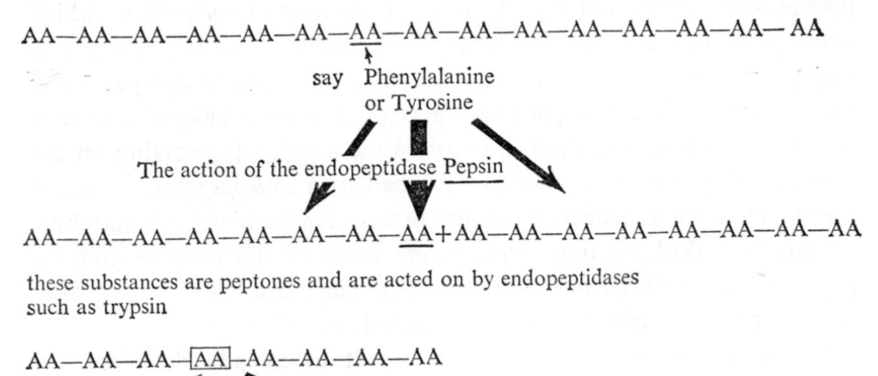

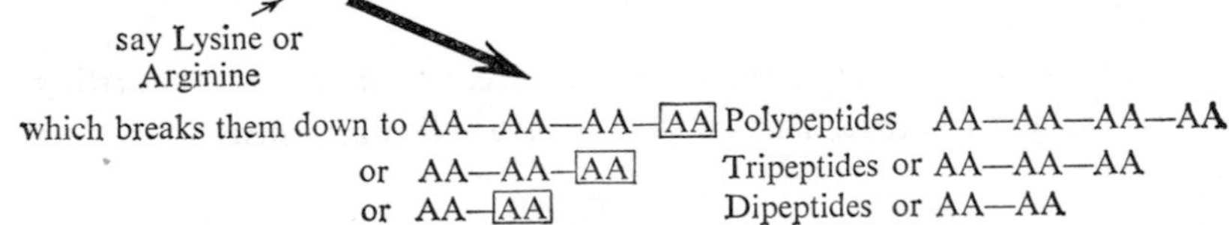

these in turn are hydrolysed by the exopeptidases, which consist of aminopeptidases and carboxypeptidases

AA—AA—AA—COOH by carboxypeptidase

AA—AA—AA—NH_2 by aminopeptidase

and finally rendered to individual amino acids

in which form they are absorbed

Fig. 12. Pathways of protein digestion in the mammal.

2.444 *Coordination of the duodenum and pancreas.* The acid which passes through the pyloric sphincter with the semi-digested chyme acts on the duodenal mucosa causing the release of the hormone *secretin*. This enters the bloodstream and eventually finds its way to the blood vessels of the pancreas causing it to secrete an alkaline fluid and thus neutralise the stomach acid. This is a good example of a feed-back system as the greater the amount of acid the more secretin and consequently the more alkali produced. A further hormone called *pancreozymin* is also formed by the duodenal mucosa and released in the presence of food. This travels to the pancreas and causes the secretion of the various enzymes already described. The pancreas is under nervous control from the vagus and this together with certain hormones controls the production of the pancreatic juices.

Bile secretion is regulated in such a way that it is released at the appropriate time into the duodenum by the contraction of the gall bladder. When food is eaten the bile starts to be released after about 1 hr and reaches its maximum flow some 2–5 hr later. Fats entering the duodenum cause production of a hormone, *cholecystokinin*, which stimulates the emptying of the gall bladder. This hormone should not be confused with secretin which besides its effects on the pancreas, also controls the rate at which the liver produces bile but not the actual emptying of the gall bladder.

Finally the duodenum itself is stimulated to secrete the succus entericus by the usual double mechanism, that is, nervous and hormonal. The vagus innervation stimulates secretion and sympathetic fibres from the solar plexus depress it, while local stimulation of the duodenal mucosa by food causes the liberation of the hormone *enterocrinin* which in turn initiates secretion by the duodenal glands. The movements of the small intestine whereby the food is pushed along towards its lower end will be described at the completion of the next section.

2.45 *The ileum*

This is the second half of the small intestine and is where the bulk of assimilation takes place. The lining epithelium secretes mucus but no enzymes and has some 20–40 villi to the square millimetre. These villi are larger than those of the duodenum (Fig. 13) and are provided with a complex network of capillaries arising from the mesenteric artery and returning into branches of the hepatic portal vein. In the centre of each villus runs a *lacteal* derived from the lymph vessels which permeate the sub-mucosa.

Some of the smaller molecules in the food, in particular glucose and galactose as well as smaller quantities of fat and amino acid, will already have been assimilated before reaching the ileum, but, as this is the main region of uptake, the various systems whereby the food actually enters the tissues of the body will be described at this point.

The mucosa of the small intestine is rich in the enzyme *phosphatase* and it seems that sugars are phosphorylated before being absorbed. Like other classes of food substance they can be assimilated against the concentration gradient and it may be that the phosphorylated sugar is able to pass across the cell membrane while the normal one is not. Once inside the cells the sugar would be released and again would not be able to escape back into the gut but could be used or transported away. Such an active assimilation requires the expenditure of energy.

Amino acids are also actively assimilated as can be shown by the fact that metabolic poisons prevent their uptake. They pass into the capillaries of the villi for transport to the liver in the *hepatic portal vein*.

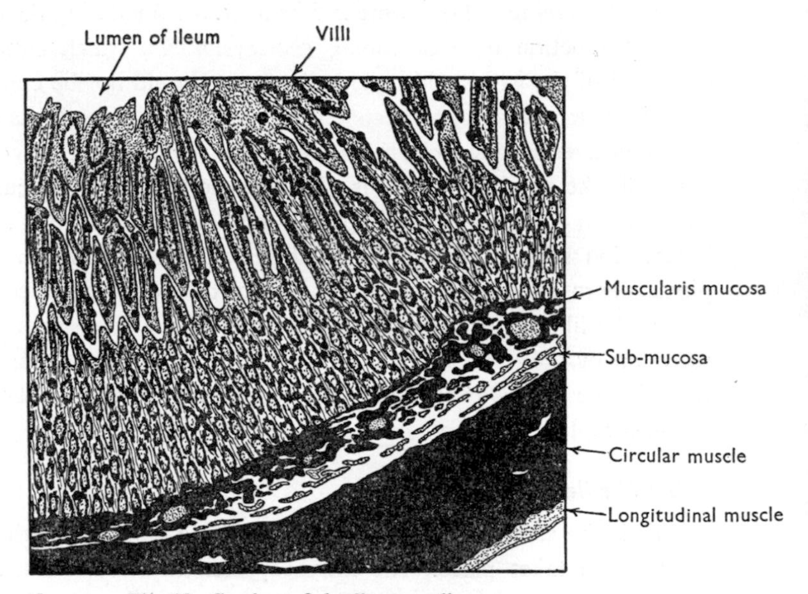

Fig. 13. Section of the ileum wall.

Fats may be assimilated as fatty acids, glycerol, glycerides, or as very small molecules of emulsified fat. The fatty acids become water-soluble in the presence of bile salts with which they combine. Some two-thirds of the fatty substances present are passed into the lacteals of the villi while the rest enters the blood capillaries. Fat that gets into the lacteal is probably unchanged fat; that which enters the blood capillaries does so as short-chain fatty acids and glycerol. Chylomicrons (i.e. small fat droplets) are formed in the gut lumen and are probably a requisite for absorption into the lymph. Their presence in the gut lumen and lymph supports the view that fat is absorbed directly.

Other substances assimilated into the small intestine are water—which is taken up osmotically as the body fluids are *hypertonic* (of higher osmotic strength) to the gut contents—vitamins and mineral salts. Vitamins pass into the gut by diffusion, the fat-soluble A and D entering with fatty substances, which is one of the reasons for requiring fat in the diet. Mineral salts are selectively absorbed: thus Na^{+} and Cl' ions enter the gut wall easily, whereas Mg^{++} and SO_4'' only very slowly.

2.46 *Movements of the small intestine*

These movements are mainly peristaltic, as already described for the oesophagus. There are, however, two other types of movement which increase the efficiency of the digestive processes in this region of the alimentary canal. These subsidiary movements are *segmenting*, which nip the food into segments and occur at about 12/min, and *pendular* movements which cause swaying of the intestine from side to side. The three systems cause a very thorough churning of the food as they pass it to and fro across the villi.

At the junction of the small intestine and the colon is the *ileo-colic sphincter*. This is normally shut and serves to prevent the forward passage of food and, more importantly, its backwards return from the colon along with harmful micro-organisms. About two hours after a meal is eaten the sphincter is reflexly opened and the contents of the gut move on, by peristalsis, into the large intestine.

2.47 *The large intestine*

A tube of wide diameter which, in man, is seen to be made up of three portions, an ascending, transverse and descending limb. Its functions are very variable in mammals but in man its importance is between that of herbivores, where it plays a vital part in digestion, and carnivores, where it is very reduced. The structure of the large intestine is as for the rest of the gut but it has no villi and no secretions other than mucus.

In man the functions of the colon or large intestine are to receive excretory substances removed from the blood, in particular Ca^{++} Mg^{++} Fe^{++} and PO_4''', and to assimilate water and other substances. Some 500 c.c. of water pass along the large intestine in 24 hr but in the same period only 100 c.c. is eliminated in the faeces. The large intestine has a rich bacterial flora and these live on undigested and indigestible residues entering from the ileum. Prominent species include varieties of staphylococci, *Escherichia coli*, *B. pyrocyaneus*, etc. Many of the bacteria are potential pathogens and if the replacement mechanism of the gut breaks down, or the host dies, they will attack the tissues of the host.

Recently it has been observed that prolonged treatment by antibiotics causes vitamin, especially vitamin B, deficiency diseases, so that it would appear that in our own case a certain amount of our vitamins are provided by our intestinal flora.

One estimate of the number of bacteria in the human large intestine is $128,000 \times 10^6$, and their bodies, mostly dead and alive, make up about 40 % of the weight of the faeces.

The descending portion of the large intestine leads into the rectum and this opens to the exterior by means of the anus.

2.48 *Movements of the large intestine and elimination of faeces*

The large intestine receives parasympathetic stimulation from the vagus and sympathetic innervation from the ganglia of the lumbar and sacral regions of the spinal cord. The movement of the contents of the large intestine takes place by mass peristalsis which is intermittent and caused by the taking in of food, or movement higher up in the gut. These movements automatically fill the rectum and when the pressure of the contents increases beyond some 40 mm Hg the anal sphincter will send stimuli to the central nervous system which leads to the elimination of the faeces.

Other than the bacteria voided the waste matter consists of indigestible matter such as cellulose which has passed unchanged through the gut and is called *roughage*. It is beneficial to include a certain amount of such matter in the diet as it gives the intestinal muscles work to do and thus maintains their tone. Other materials include dead cells from the gut lining and bile pigments.

2.5 Modifications of the alimentary canal found in mammals other than man

2.51 *The teeth and jaws*

In a mammal these indicate quite clearly the feeding habits of their owner. As already described for man the teeth of mammals fall into four categories. In the front of the mouth are the chisel-shaped incisors, which are followed by a single sharp canine, after this are the premolars which may be modified for grinding but do not usually show the complex cusp patterns of the back teeth or molars.

2.511 *Herbivores.* In herbivores the incisors are large and continuously self-sharpening by having a leading edge of *enamel* working against softer *dentine* in the opposing tooth. The canines are usually absent and in their place a large gap or diastema exists which allows the food to be freely circulated in the mouth. The premolars tend to become *molariform* and work in conjunction with the molars, the two types of teeth making up a broad grinding surface. Like the incisors, these latter teeth show continuous growth and also the most specialised of mammalian cusp patterns. In order to achieve a close-fitting and indented surface the cusps are made up of an arrangement of enamel hills and dentine valleys placed in such a way that those of the one jaw fit into those of the other. Owing to the different hardnesses of these two substances an excellent grinding surface is maintained throughout the life of the animal. Besides the cusps of the individual teeth the whole system falls into a series of ridges running either transversely or in the line of the jaw. The direction of these ridges will conform with the direction of movement of the jaw.

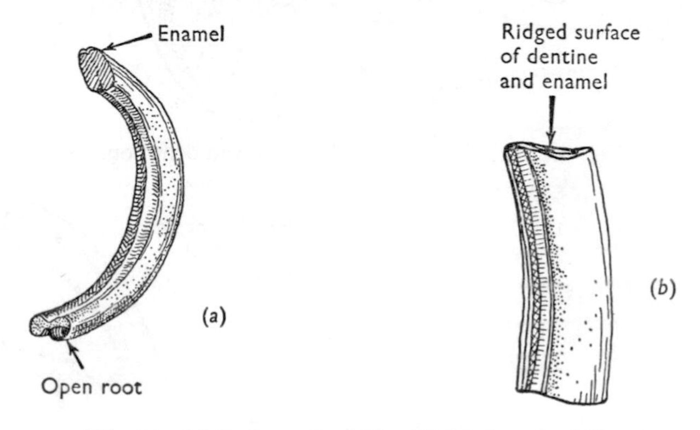

Fig. 14. (*a*) Incisor of rabbit. (*b*) Molar of rabbit.

As a herbivorous adaptation the importance of close grinding, which reduces the food to a cellular level, cannot be overemphasised. In the first place the food is rendered into a form in which it is available for rapid bacterial action in the rumen or intestine, and in the second place destruction of the cell structure leads to processes of *autolysis* or self-digestion. By this process the complex proteins and other chemicals of the protoplasm are destroyed by the activity of their own enzymes.

Together with the specialised teeth, the mandible of the herbivore shows several adaptations towards an efficient chewing action. For example the coronoid process is much elongated and this gives the masseters a large moment about the articulation. The temporal on the other hand has little leverage and is much reduced (Fig. 16). The chewing action is performed by the masseters rolling the narrow lower jaw from side to side

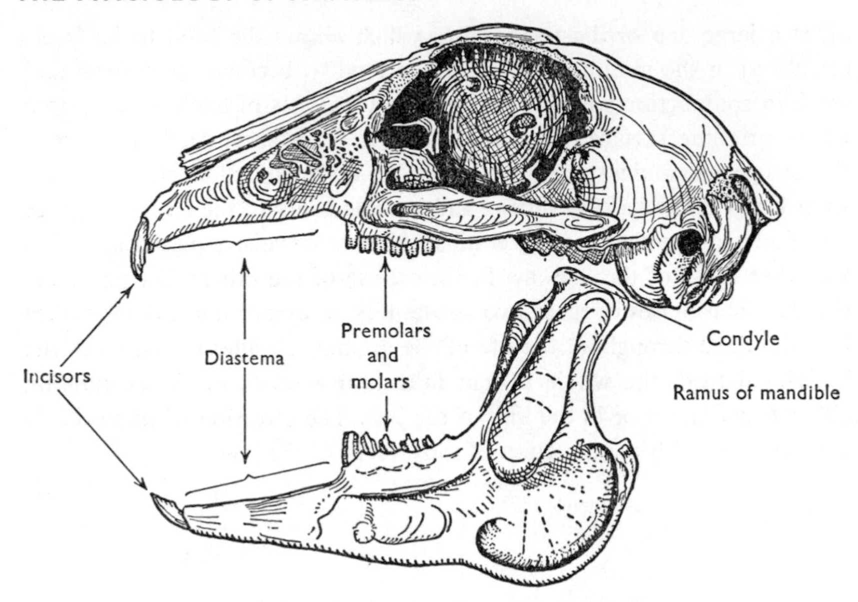

Fig. 15. Rabbit: showing diastema and dentition.

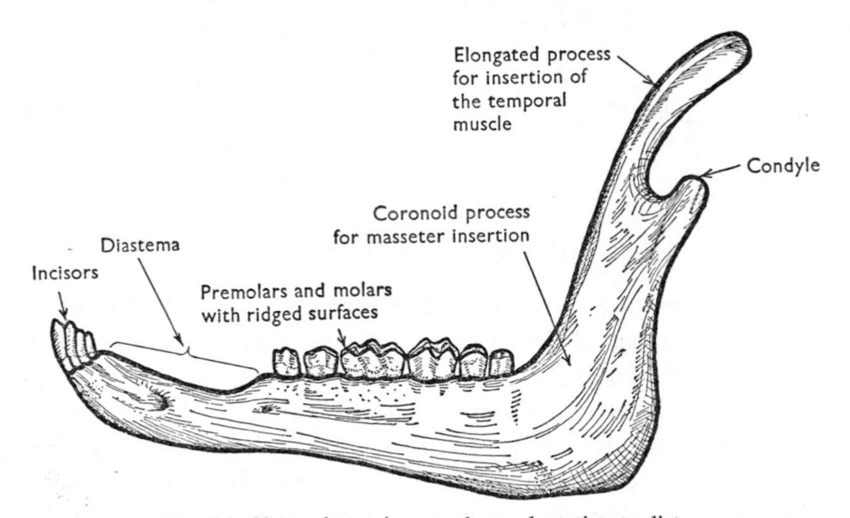

Fig. 16. Sheep: lower jaw, to show adaptation to diet.

against the upper. There is little vertical force on the mandible and it is weakly constructed compared with those of carnivores.

The joint that the posterior condyle makes with the glenoid is loose and allows the lower jaw to move freely from side to side. This is only possible in a system which uses small vertical forces.

The dental formula of three typical herbivores is as follows:

Rabbit	$I\frac{2}{1}C\frac{0}{0}P.M.\frac{3}{2}M\frac{3}{3} = 28$
Horse	$I\frac{3}{3}C\frac{1}{1}♂P.M.\frac{3}{3}M\frac{3}{3} = 40$
Sheep	$I\frac{0}{3}C\frac{0}{1}P.M.\frac{3}{3}M\frac{3}{3} = 32.$

2.512 *Carnivores.* In carnivores all the teeth find some use in feeding and, on the whole, there tend to be more of them than in the previous types described. In the front of the mouth the incisors are not particularly well developed though they find some use in grooming and nibbling meat from bones. Behind the incisors are a pair of very large canines in each jaw and these stabbing teeth are used to kill the prey. The premolars and molars are covered in enamel and rise to sharp edges along the line of the jaw, the

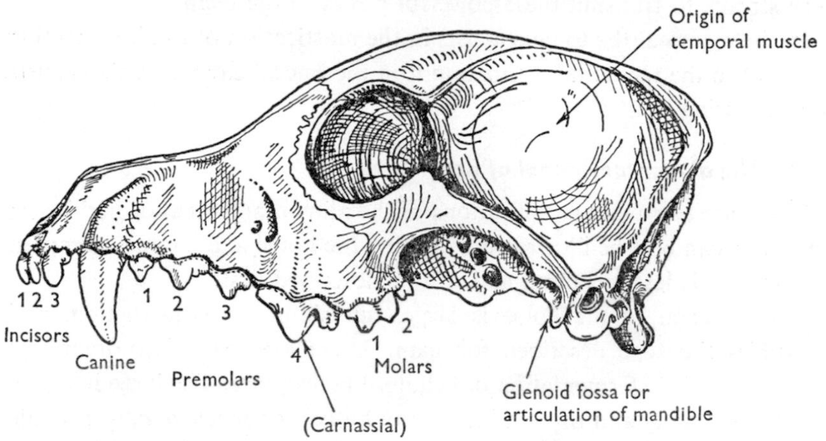

Fig. 17. Dog skull: adaptation for a carnivorous diet.

joint action of upper and lower teeth providing a powerful shearing action like scissors. Further adaptation of these teeth includes an enlarged 4th premolar in the top jaw working against a similarly enlarged 1st molar in the bottom jaw. These teeth, adapted for cutting and bone crunching, are called the *carnassials* and are found in the cat and other families of carnivores.

Thus the dentition of carnivores is more adapted to tearing than to grinding as flesh does not require the same degree of mastication as plant tissues. For this reason, though the teeth are large and very strong, they do not have the same wear as those of herbivores and are not open-rooted.

The articulating condyle of the dentary is placed low down and consequently the forces exerted by the temporal and masseter are more evenly distributed about the joint. Part of the strain on the jaw is provided by the forward and downward struggles of the prey and these are balanced by the opposite force of the large temporal. The other main strain involved in bone crunching and tearing meat is taken by the masseter, which acts perpendicularly to the line of the teeth.

The joint that the lower jaw makes with the glenoid is a tight roller joint and this is as important to the efficient working of the jaw as is a tight rivet to scissors. Held by this joint the jaw moves only in the vertical plane. A further point is that the distribution of the temporal and masseter at equal distances from the point of articulation enables them to take up the forces developed during the crunching of bones, and prevents these being carried to the joint itself which could cause dislocation. The dentary is very strong, to transmit these powerful forces to the teeth.

In the mammal the tongue assists in the mastication of food, presenting material to the teeth and circulating it in the buccal cavity. It also assists in swallowing.

2.52 *The alimentary canal of herbivores*

There are two main modifications of the alimentary canal of herbivores and these can best be illustrated by taking the horse and cow as examples. The former is less advanced and will be described first.

The stomach of the horse is large, but simple in construction, and resembles the type described for man. There are few micro-organisms present and little fermentation of cellulose takes place. The large intestine is extremely long and thrown into many bulges, or *sacculated*, to provide a stagnant environment for the activity of micro-organisms. These are able to break down cellulose and release simple food substances into the lumen of the gut whence they can be assimilated. The breakdown of bacterial protoplasm also provides vitamins. Thus the micro-organisms have come to occupy an essential and symbiotic relationship with their mammalian hosts and open up a wide range of foodstuffs which would not be digestible without their aid. On the other hand their activity below the region of assimilation (that is, the ileum) is not very efficient as the large intestine has only limited powers of absorption. In some herbivores of this type, such as the rabbit, the products of bacterial digestion in the large intestine or the appendix are passed out of the anus and eaten again so that nutritive matter can be assimilated. (This habit is called coprophagy.)

A more advanced herbivorous adaptation is found in certain ruminants, such as the cow. These have very elaborate stomachs divided into four regions (Fig. 18). The first of these regions is a large *rumen* and into this food passes from the oesophagus. In the rumen it is churned and returned to the buccal cavity for further chewing; this corresponds to the 'chewing the cud' phase of digestion. If fine, the food may pass directly into the *reticulum* or second compartment of the stomach, but the major fermentation takes place in the rumen and the food may remain here for as long as seven days. After passage through the rumen and reticulum the food enters the *omasum* where surplus moisture is absorbed or squeezed out,

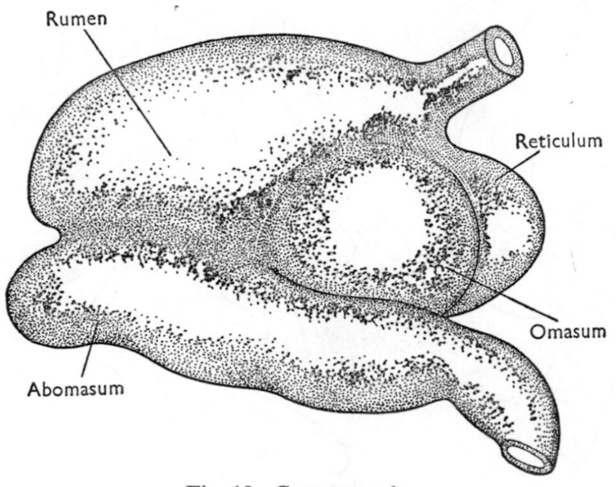

Fig. 18. Cow stomach.

this part of the stomach being particularly muscular. Finally the semi-digested food passes to the *abomasum*, the last region of the stomach, where the normal complement of enzymes operate. It is interesting to note that the epithelia of the first three regions of the ruminant stomach are cornified and tough like the lining of the oesophagus and provide a rough surface for mechanical breakdown. They are not impermeable, however, and many of the products of micro-organism digestion pass into the blood circulating the rumen. This is particularly true of fatty acids.

Contrasting with the stomach, the large intestine of ruminants is simple, though long, and little fermentation takes place in it. Despite the complex anatomy of the former an *oesophageal groove* exists whereby certain types of food can be passed rapidly along the wall of the stomach from oesophagus to duodenum.

2.6 The alimentary canal in other vertebrates

2.61 *Dogfish*

Unlike the mammal, the teeth of the dogfish are all the same and are derived from the placoid epidermal scales which they much resemble. The oesophagus is short and a sphincter prevents the entry of water into the gut during normal gill ventilation. The lobed stomach has cardiac and pyloric portions well differentiated and there is a short duodenum into

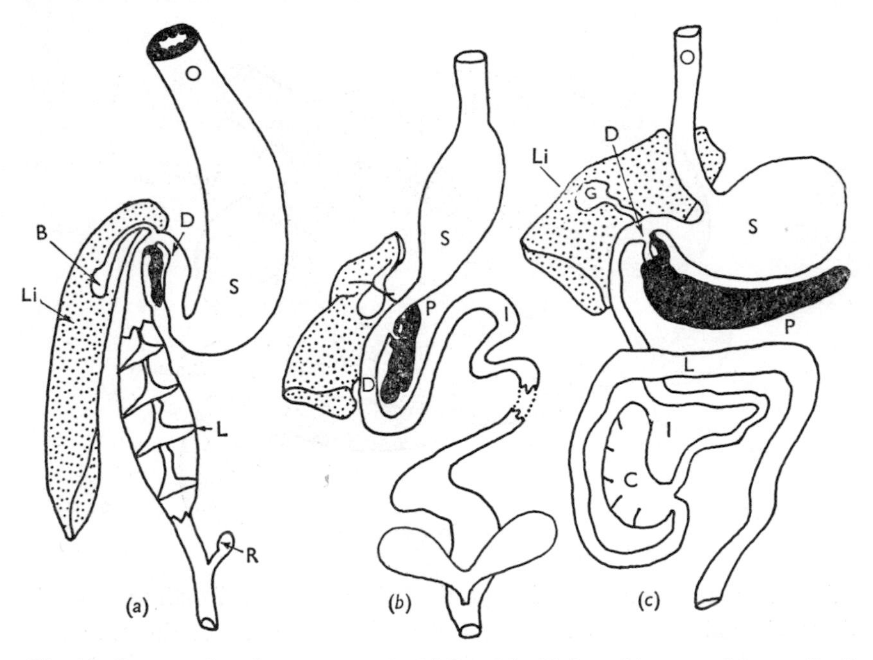

Fig. 19. Comparative alimentary canals. (*a*) Dogfish, (*b*) frog, (*c*) mammal (generalised). O: oesophagus, S: stomach, D: duodenum, P: pancreas, B: bile duct, Li: liver, R: rectal gland, I: ileum, C: caecum, L: large intestine, G: gall bladder.

which open the pancreatic and bile ducts. This is followed by the intestine, which is thick-walled and has a spiral valve passing down its centre. Finally the intestine leads to the anus. The bulk of digestion is completed in the stomach and assimilation takes place in the valvular intestine, which, with its spiral modification, has a very large internal surface. The rectal gland has a function in control of ionic equilibrium, particularly in the excretion of sodium chloride.

2.62 *Frog*

The frog has vestigial teeth and catches its prey by a distensible tongue attached to the front of the mouth. Unlike the dogfish its oesophagus is long and glandular and leads to a stomach, which despite its poor anatomical differentiation is very rich in glands and produces in the crypts of its walls a powerfully acid secretion. Following the stomach is the duodenum with secretory and mucus cells, and into this bile and pancreatic ducts open. The frog does not possess the crypts of Lieberkühn found in the mammalian duodenum. Finally the ileum has special absorptive cells but no villi and leads into the short and thick-walled large intestine, which terminates in a rectum which opens into a cloacum.

2.63 *Bird*

The great divergence of the beak in birds according to their diet parallels the teeth of the mammal. Like the beak, the *crop* which ends the oesophagus may be variable, but generally its function is to sort, and if necessary reject, substances taken in at the mouth. An interesting adaptation of the crop occurs in the pigeon where a milk-like fluid is produced by the sloughing off of the lining epithelium and this is used to initiate feeding in the young.

Following the crop is a muscular *gizzard* which may contain stones and serves to grind up the food. This adaptation fulfils the function of the teeth in other vertebrates and opens up a wide range of foodstuffs. The remainder of the gut is as already described for other vertebrates except that no gall bladder is present.

RESPIRATION

Respiration is vital to all living organisms because it provides the energy required not only to maintain their body structure but also to carry out their varied activities. The energy is liberated by oxidative processes in the cells and for this purpose the oxygen is taken in at a respiratory surface where carbon dioxide is usually liberated. It is convenient, therefore, to separate *external* from *tissue* or cellular respiration, but in animals the size of vertebrates a third *transport* process is required. In very small organisms, external respiration is mainly by diffusion across the external membranes to the sites of oxidation. Knowing the rates of oxygen consumption and diffusion it has been estimated that diffusion alone can supply sufficient oxygen only if the organism is less than 1 mm in diameter. Above this size it is *essential* that transport mechanisms are present to deliver the oxygen from the respiratory surface to the cells where it is utilised. This is because the external surface becomes relatively smaller, as it only increases as the square of the linear dimension whereas the volume using the oxygen increases as the cube.

The volume of oxygen required by an animal is often expressed as the number of c.c. per kilogram per hour (see p. 12). Animals vary considerably in their oxygen requirements, mammals and birds using larger amounts than the lower vertebrates. This generalisation only holds for animals of the same size, because in all groups there is an inverse relationship between body size and metabolic rate. For example, a small mammal such as a shrew may require 10–20 times more oxygen per unit weight than a man. Mammals and birds require a continuous supply of oxygen if they are to maintain their metabolic rate and high body temperature. In some cases this high level of metabolism is interrupted and the animal passes into a state of hibernation for a long period. Some small mammals and birds undergo similar reductions in temperature and metabolism which may occur in a 24-hour rhythm. For example, humming birds and bats enter a so-called 'torpid state' during the night and day respectively.

In both aquatic and terrestrial vertebrates the final path of oxygen from the medium to the blood is by diffusion across an aqueous phase covering the respiratory epithelium. Diffusion in air is rapid but in water only at $\frac{1}{300,000}$ times the speed. There is a danger, greater in water than air, that stagnant layers may form at the surface with consequent reduction in the rate of gaseous exchange. Animals partly overcome this danger by rhythmic ventilation of the respiratory organ. In water the gill is a finely divided external expansion of the surface epithelium which is readily ventilated and is supported by the aqueous medium. The dangers of water loss are absent, but on land such an organ would be useless. On land respiratory organs are usually in-tuckings of the surface and the small

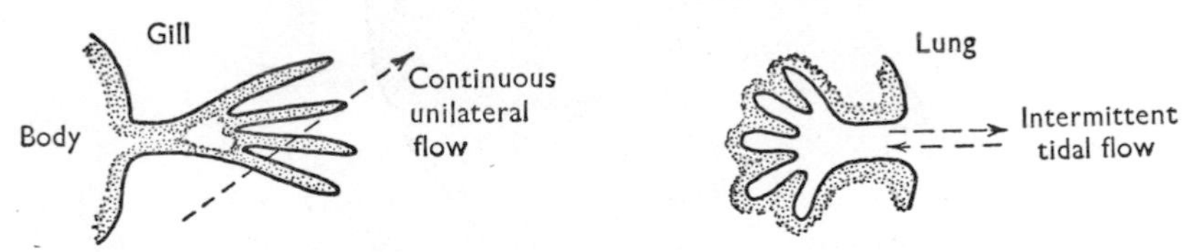

Fig. 20. The gill and lung in relation to the body and respiratory medium.

proportion of the moist epithelium directly in contact with the external air enables gaseous exchange to take place with the minimum loss of water. Ventilation of such an organ is more difficult than of a gill where a continuous flow of water is readily achieved, but in most lungs the ventilation of the respiratory epithelium is tidal. The lungs of lower vertebrates are simple sacs and as we go up the vertebrate series an increase in the folding of the respiratory epithelium is found. In all vertebrates there is a relationship between the size of the respiratory surface and the activity of the organism (Table 3, page 59).

3.1 Respiration in mammals

3.11 *The respiratory tract*

The lungs are contained within the thorax where they are surrounded by the inner and external *pleural membranes*, mesodermal in orgin, which are normally closely apposed to one another. In certain diseased conditions, for example, pleurisy, they may become inflamed, or the pleural space may be filled with liquid or air.

Air enters the respiratory tract through the nostrils which act as filters and warm the air before it passes above the soft palate to the back of the throat on its way to the lung. Posteriorly the nasal passage enters the *pharynx* into which also opens the cavity of the mouth. Leading off from the pharyngeal cavity are several passages. The most important of these are the *oesophagus* down which the food passes, and ventral to this is the opening to the respiratory tract proper. This opening is protected by the *epiglottis*, which prevents food entering the larynx during swallowing. The

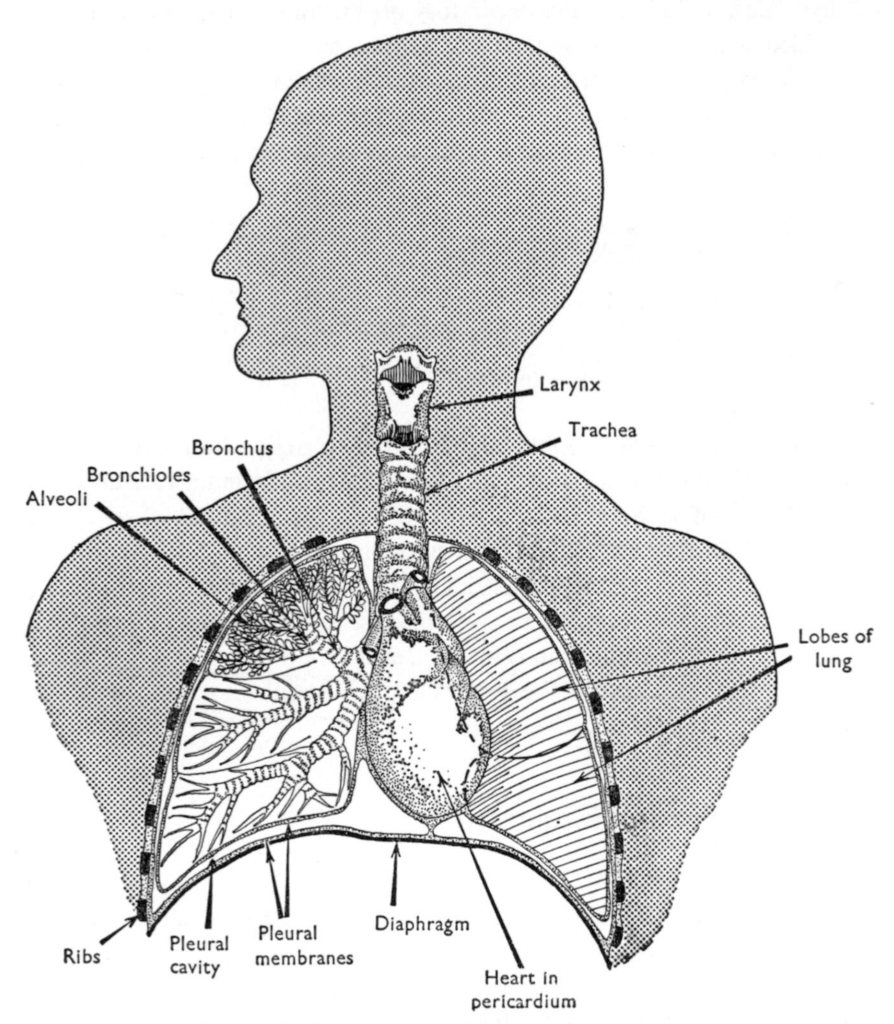

Fig. 21. The human lung and associated structures.

epiglottis projects upwards from the floor of the pharynx and is stiffened by an elastic cartilage. The *larynx* is supported by several cartilages and internally contains the vocal cords, which may be vibrated by the air stream. From the pharynx the *trachea* leads posteriorly and bifurcates into the two primary *bronchi*. All these tubes are supported by incomplete rings of cartilage which prevent their collapse when the pressure inside is reduced. The rings of cartilage are reduced to irregular plates in the smaller tubes and are completely absent from the *bronchioles*. The bronchioles branch repeatedly until they finally end in alveolar sacs, or air sacs, the walls of which are minutely sacculated and form the *alveoli* where gaseous exchange mainly occurs. There may be as many as 350 million alveoli in each lung of a man. Each sac is about 100 μ across and its surface is extremely thin (less than $\frac{1}{2}\mu$) and is closely associated with the endothelium of the blood capillaries. The total internal respiratory surface of this structure is immense. Because of its extensive nature, it is not unexpected that the properties of this surface are vitally important in the functioning of the lung. Recently, attention has been drawn to the role of surface-tension forces in the outermost molecular layer of the thin film of fluid that lines the alveoli. These forces produce up to three-quarters of the total elasticity of the lung, which must be overcome each time the lungs are inflated. Certain cells in the alveolar lining secrete a 'surface-active' agent which reduces this surface tension and prevents the lung from collapsing. Absence of this detergent-like substance in new-born babies is the cause of a disease which may be fatal.

The lung forms a quite compact structure as it is supported by the presence of a connective tissue with many elastic fibres. Many of the conducting tubes contain a muscular layer which increases down to the terminal bronchioles where it is thickest in proportion to the other layers. But in the still finer branches of the bronchial tree elastic tissue is prevalent as a sort of mesh-work through which the alveoli are evaginated. The epithelium is ciliated and changes progressively from a columnar type in the larger tubes to a pavement epithelium in the alveoli.

3.12 *Ventilation*

The elastic lungs are contained within the thoracic cavity, which is a closed box supported externally by the ribs and separated posteriorly from the abdominal cavity by the dome-shaped *diaphragm*. Changes in volume of the thorax immediately cause slight rarefaction or compression of the air in the lung. These changes in pressure result in the flow of air into and out of the lungs (Fig. 22). The pressure within the thoracic cavity (intra-thoracic pressure) is normally about 4 mm Hg less than atmospheric, which is the pressure in the lung when at rest. This difference in pressure is probably due to changes which occur during development. In the early stages the lungs completely fill the thoracic cavity but later the thorax increases in volume to a greater extent than the lung. This results in the lungs becoming stretched away from the thoracic cage. Consequently a reduced pressure within the thorax is produced. During inspiration the *intrathoracic* pressure is reduced by some 5 mm Hg and this reduction is transmitted to the lung so that air flows in to restore the *intrapulmonic* pressure to atmospheric. During expiration the intrathoracic pressure rises again to its resting level and the increase in pressure compresses the air in the lung causing its pressure to rise above atmospheric. Air is forced

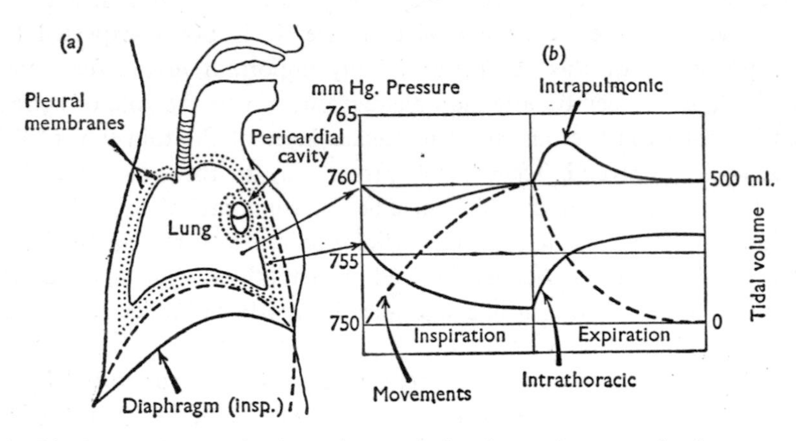

Fig. 22. Pressure and volume changes during the respiratory cycle of man.

out until once more the pressure within the lung is restored to atmospheric before the start of the next inspiration.

Changes in volume of the thoracic cavity are brought about by two main types of muscular action. The diaphragm may contract and effectively increase the antero-posterior dimensions of the thoracic cavity, or the ribs may be raised, so increasing the cross-sectional area of the thorax. These two movements are not entirely separate, for contraction of the diaphragm also has the effect of raising the costal margin. Normally a combination of *diaphragmatic* and *costal* pumping is used, but under certain circumstances sufficient ventilation can be achieved by only one of these actions. The diaphragm is dome-shaped anteriorly and is attached to the lumbar vertebrae and the posterior ribs. The central tendon is not domed upwards as much as on the two sides so that during relaxation the right and left sides of the diaphragm are elevated above the central tendon by the pressure of the abdominal contents. These arches are flattened when the diaphragm contracts and the central tendon moves down about 1·5 cm (Fig. 22). The diaphragm is innervated by paired phrenic nerves which contain the motor fibres. Section of one phrenic nerve abolishes the contraction of the diaphragm on that side.

The ribs articulate with the thoracic vertebrae by two heads, a *tuberculum* to the transverse process and a *capitulum* to the centrum. Movements of the ribs are about the line joining these two articulations. At rest, the ribs are inclined backwards and ventrally with respect to the vertebral column

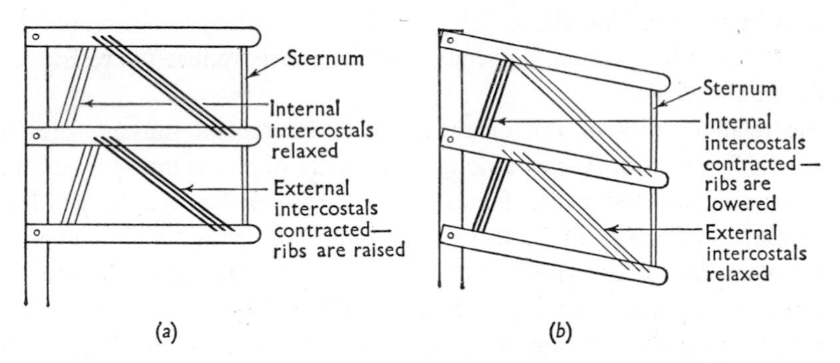

Fig. 23. The use of the antagonistic intercostal muscles in raising and lowering the rib cage.

but during inspiration they move anteriorly with a consequent increase in the transverse dimensions of the chest. In human infants, the ribs are nearly at right angles to the vertebral column and hence any movement, whether forward or backward, decreases the chest volume, consequently ventilation is largely diaphragmatic. The upward and outward movements of the ribs has been likened to the action of a bucket-handle about the articulations with the vertebral column and sternum. The *intercostal* muscles which produce these movements are divisible into external and internal groups. During quiet breathing no distinction can be made between the action of these two groups, but in more active ventilation the external group is clearly involved in the inspiratory movement (Fig. 23). This action is due to their attachment to the anterior rib being nearer to the vertebral column than the insertion on the more posterior rib. The internal intercostals have the relative positions of their insertions reversed and thus tend to lower the rib cage when they contract. The first rib and its attachment to the sternum remains fixed during quiet breathing and forms a point relative to which the other ribs move.

Expiration is mainly produced by passive relaxation of the inspiratory musculature and a return of the ribs and diaphragm to their resting position. Even in quiet breathing, however, there is evidence that it is not entirely passive because some of the inspiratory muscles remain slightly active even when the volume of the thoracic cavity is increasing. This leads to a more controlled outflow of air from the lung. During vigorous breathing active expiratory muscles are brought into action, mainly those of the abdomen, which increase the pressure within the abdominal cavity and help to force air out of the thorax. The internal intercostal muscles also assist in this. The size of the nostrils and glottis are increased by muscles during inspiration, movements which clearly reduce the resistance to the inflow of air.

The relative roles of the diaphragm and intercostal musculature in producing ventilation varies among individuals of the same species and between different mammals. Human infants rely on their diaphragm, but among adults females make use of the intercostal muscles to a greater extent than do males, who breathe more diaphragmatically. Mammals which normally stand on four legs also use the diaphragm to a greater extent, because the rib cage is involved in a supporting function. Correspondingly, amongst aquatic mammals the body is supported by the medium and the intercostal musculature is more developed. The diaphragm of many diving mammals is more horizontal in the body and facilitates the transmission of pressure changes in the external medium to the pleural cavities. Whales have relatively small lung volumes (2·5 c.c./100 g body weight) and a large proportion of dead space. In seals the lungs are about 5 c.c./100 g body weight which is similar to the figure for a man. The tidal air of divers makes up a large proportion of the total lung volume (80 % in a porpoise) and assists in the rapid exchange of gases when they surface.

During normal breathing the respiratory movements occur about 32 times per minute in a new-born child, 26 times in a child of five, 16 times in a man of 25, and 18 times in a man of 50.

In man, the total capacity of the lungs is about 5½ litres. But even following the most forced expiration 1½ litres or more of this cannot be forced out of the lung. The maximum volume which can be ventilated during forced breathing is referred to as the *vital capacity* of the lungs. This is normally between 3 and 4 litres and may be as high as 5 or 6 litres among athletes. The volume of air moving in and out of the lung during normal breathing makes up the *tidal volume*. During quiet breathing a man takes in about 500 c.c. of air. Of this volume only a certain part reaches the alveoli as the rest fills the air tubes, etc., leading to the respiratory epithelium. This forms the *dead space* and normally amounts to 140 c.c. of the tidal volume. Consequently the volume actually reaching the alveoli (*the alveolar air*) is little more than 360 c.c. Because ventilation is tidal, during expiration the dead space is filled with air of the same composition as alveolar air and by forced expiration it may be sampled.

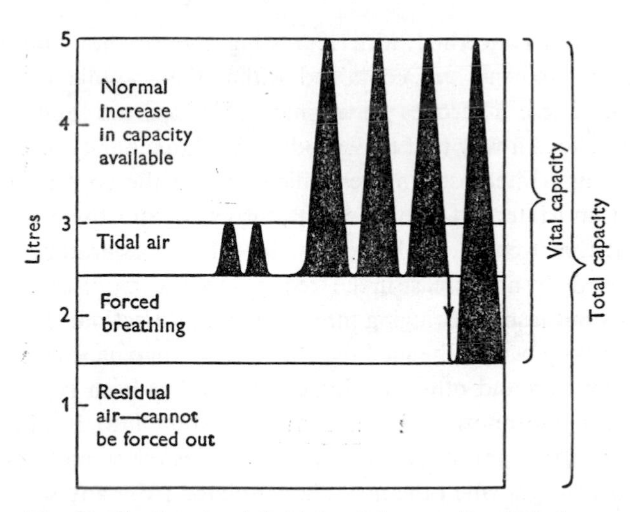

Fig. 24. The functional divisions of the capacity of the lung.

The composition of the alveolar air at rest and under normal atmospheric conditions remains very constant, in fact the whole of the regulatory mechanisms of respiration are designed to maintain the constancy of this air. The compositions of the inspired and expired air differ mainly in the proportion of oxygen and carbon dioxide. *Inspired air* contains 21 % of oxygen and 0·04 % of carbon dioxide, whereas the *expired air* has only 16·4 % oxygen but 4·1 % carbon dioxide. The percentage of carbon dioxide increases when the tidal volume is increased because the dilution effect of the dead-space air diminishes. The marked difference in composition of the inspired and expired air contrasts with the relative constancy of the alveolar air. This is partly because the tidal air normally forms a small proportion of the total air in the alveoli. Thus at the end of normal expiration the alveoli still contain $2\frac{1}{2}$ litres of gases. In inspiration, 360 c.c. of atmospheric air are taken into this space and mixed with the $2\frac{1}{2}$ litres already present. Consequently the effects of this small ventilation of the total lung air are very slight and will only amount to less than ½ %. Haldane determined the alveolar CO_2 at the end of inspiration and expiration respectively; his figures were 5·54 % and 5·70 %. Thus the alveolar air can function as a sort of buffer between the external air and gas tensions of the blood.

3.2 The nervous coordination of respiration

The neurones concerned with regulating the rhythm and depth of respiratory movements are contained within the medulla oblongata of vertebrates. These neurones have intrinsic properties and are interconnected in such a way that they produce rhythmic bursts of impulses in the phrenic and intercostal nerves which lead to the contraction of the diaphragm and intercostal muscles. Recordings from these nerves have shown that the strength of the contraction of the respiratory muscles is regulated partly by an increase in the frequency of the impulses to individual motor units but also by bringing more of them into action. Recording of electrical activity within the medulla shows that some neurones are active during inspiration and others during expiration but both types are intermingled with one another, i.e. a clear separation into inspiratory and expiratory centres cannot be made. The neurones continue to show their rhythmic discharges when the medulla is isolated from any sensory input either by cutting all the cranial nerves or by abolishing the respiratory movements by the injection of curare (Indian arrowhead poison) which paralyses the muscles. The basic respiratory rhythm is therefore intrinsic to these neuronal networks; but it can be modified by inputs from proprioceptive sense organs in the bronchial tree. During normal respiration, discharges from these sense organs can be recorded from the vagal nerves, and they reflexly initiate the next phase (inspiration) of the cycle. The medullary neurones are readily influenced by other neurones in the brain, notably in the regions near the *pons* (p. 219).

The rate and depth of ventilation are precisely regulated according to the composition of the alveolar air. Small increases in CO_2 (0·25 %) may double the ventilation volume. This effect is mainly due to the increased CO_2 tension of the blood affecting the medullary regions of the brain. Cells here are extremely sensitive to CO_2 and immediately alter the respiratory rhythm. Injection of CO_2 in bicarbonate buffer produces effects similar to those produced by heightened CO_2 tensions of the alveolar air. The CO_2 tension is also detected by receptors in the carotid body and aortic arch. These regions are most sensitive, however, to changes in the O_2 tension, which again has some controlling effect, though not so great as the CO_2 tension. The respiratory centres of the medulla are very insensitive to lowering of the O_2 tension.

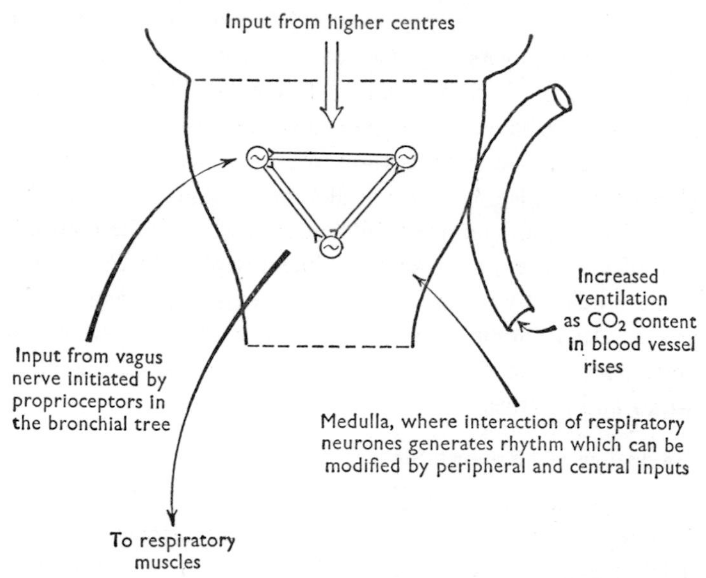

Fig. 25. Generation of the respiratory rhythm in the mammalian medulla.

In addition to these factors, which are clearly respiratory in nature, the respiratory rhythm may also be influenced by effects produced in other parts of the central nervous system. For instance, stimulation of certain areas within the cerebral hemispheres or of the cerebellum may affect the respiratory rhythm. It is an everyday experience that the respiratory movements can be controlled voluntarily.

3.3 Transport of the respiratory gases

The source of oxygen is one factor which limits the rate at which it can be obtained by an animal. One litre of air contains approximately 200 c.c. of oxygen, whereas natural waters contain between 0·05 and 9 c.c. of oxygen per litre depending on the type of water and the temperature.[1]

[1] The amount of oxygen which will dissolve in a solution is determined by its solubility and partial pressure in the gas mixture. The value of the partial pressure of the gas with which the solution is in equilibrium is referred to as the tension of the gas in the

With very few exceptions (e.g. Leptocephalus larva of eels and some antarctic fish), the oxygen-carrying capacity of vertebrate blood is increased by the presence of haemoglobin. This respiratory pigment is contained within corpuscles which enable the blood viscosity and osmotic pressure to be kept smaller than if the same amount were dissolved in the plasma.

The study of the functioning of haemoglobin as a respiratory pigment has been greatly aided by its very characteristic absorption spectrum under different conditions. Oxyhaemoglobin has two chief bands, whereas reduced haemoglobin has a single broad band in the yellow-green. Haemoglobins readily combine with carbon monoxide to produce carboxyhaemoglobin which also has two well-defined bands. Carboxyhaemoglobin is unable to combine reversibly with O_2, which is the reason for its poisonous effects. The importance of haemoglobin to an animal can therefore be studied by following the effects of CO poisoning, which in most cases is lethal. Some fishes, however, are able to survive when all their haemoglobin is rendered useless, so long as they are kept at low temperatures and inactive.

3.31 *The nature of haemoglobin*

Haemoglobins consist of an iron-porphyrin prosthetic group—*haem*, and a protein portion—*globin*. The haemoglobin molecule is made up of two equal halves (molecular weight = 34,000) each of two polypeptide chains (α and β) which together are composed of 290 amino acids. There are four haem groups, each associated with one of the polypeptide chains. Differences in the properties of haemoglobins are due to small changes in the polypeptide chains. Foetal haemoglobin differs from adult haemoglobin only in the polypeptide chains. The species-specific properties of oxygenation are probably due to small changes in the chains at the regions where they coil in the vicinity of the iron atom. The very marked effects of small changes in these chains have been well worked out for sickle-cell anaemia. The hundredfold reduction in solubility of the haemoglobin is here due to the place of one glutamic-acid amino acid, out of a total of 290 residues, being taken by a valine residue.

solution. For example, water saturated with oxygen corresponds to a tension of 152 mm Hg, which is the partial pressure of oxygen in air. This tension will be reduced if the partial pressure of oxygen in the air falls, as for example at high altitudes. It is customary to refer to the oxygen in terms of its tension when considering diffusion and uptake at the respiratory epithelium because the equations relating to these phenomena are expressed as partial pressures. On the other hand, when considering the amount of oxygen available to an airbreathing or aquatic animal the volume of gas per litre is a more useful way of expressing the concentration.

3.32 *Combination with oxygen*

An important property of any blood pigment is to increase the volume of oxygen carried by a given volume of blood when fully saturated (Table 3). For most birds and mammals it is usually between 15 and 20 vol. %, being higher in some diving forms, for example, seals (29 vol. %), but lower in most amphibians and fish. The combination between haemoglobin and oxygen is a reversible process which varies according to the oxygen tension with which the blood is in equilibrium. The oxygen combines loosely with the ferrous iron of the haemoglobin in the proportion of 1 molecule per atom of iron. Each molecule of haemoglobin can therefore combine with 4 molecules of oxygen:

$$Hb + 4O_2 \leftrightharpoons HbO_8.$$

Table 3. *Oxygen consumption of various vertebrates*

Animal	O_2 requirements (c.c./kg/hr)	Lung or gill surface area (sq.cm/g)	O_2 carrying capacity of blood (vols. %)
Man	330	7	20
Mouse	2500–20,000	54	*c.* 17
Sparrow	6700	—	*c.* 16
Frog	150	2·5	10
Dogfish	54·5	18·6	5

The curve for such an equilibrium would have an extremely marked inflexion but the peculiar sigmoid (S-shaped) shape of the oxyhaemoglobin dissociation curve is explained by the intermediate compound theory:

$$Hb + O_2 \leftrightharpoons HbO_2$$
$$HbO_2 + O_2 \leftrightharpoons HbO_4$$
$$HbO_4 + O_2 \leftrightharpoons HbO_6$$
$$HbO_6 + O_2 \leftrightharpoons HbO_8.$$

Throughout this remarkable but little-understood process of oxygenation, the iron maintains its ferrous condition.

Haemoglobin functions by giving up oxygen at low tensions and combining with it at high tensions. The relationship between tension and percentage saturation of the blood is not linear, however, as can be seen from the dissociation curve (Fig. 26). Most haemoglobins become fully saturated at oxygen tensions below that at their respiratory surfaces. This leads to a 'safety factor' whereby the ambient oxygen tension can be lowered quite considerably before the blood is no longer saturated with oxygen. The steep part of the sigmoid curve usually coincides with the normal range of tensions over which it functions in the animal. Some indication of the position of the curve is given by reference to the *unloading tension* (Tu), sometimes called P_{50} because it is the tension at which the haemoglobin is 50 % saturated. The higher the value of P_{50} maintained in the animal the better, as it gives some indication of the level of oxygen tension in the tissues. Birds have high P_{50}'s (duck, 50 mm Hg), in mammals it is about 30 mm Hg and for fishes the value is in the range 10–20 mm Hg. The transfer of oxygen from the maternal to foetal circulation is aided by differences in the dissociation curves of the two haemoglobins. The foetal curve is to the left of the maternal curve, so that it becomes oxygenated at lower partial pressures (Fig. 26). There is also a counter-flow between foetal and maternal bloods in the placenta of some mammals (Fig. 48).

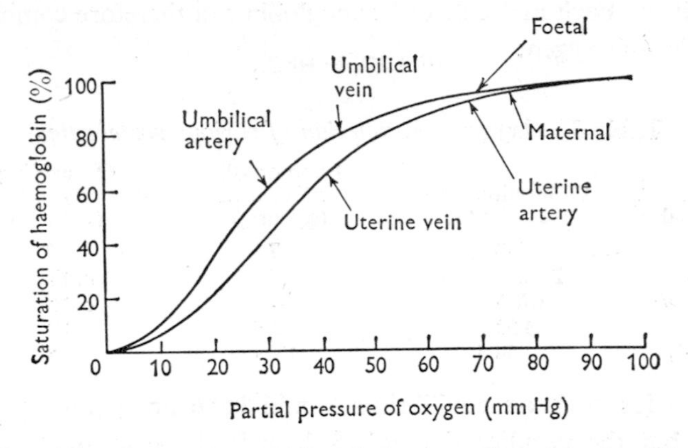

Fig. 26. Comparison between oxygen dissociation curve for maternal and foetal blood. The foetal blood (with its curve to the left of the maternal) will load up with oxygen from the mother's blood at a given partial pressure. The data given is for man.

The dissociation curve of bloods varies according to the CO_2 tension. At higher CO_2 concentrations the curve is moved to the right—the so-called 'Bohr shift'. This has the important effect of increasing the amount of oxygen liberated from the haemoglobin in tissues where the CO_2 content is high (Fig. 27). The actual curve along which the blood functions in the body in relation to oxygen lies between the curves determined experimentally at different CO_2 tensions. This effective curve is steeper than those at a single CO_2 tension and consequently leads to a greater liberation of oxygen for a given difference in O_2 tension between the lungs and tissues. The haemoglobins of tetrapods and fishes which live under conditions of high CO_2 and low O_2 tension tend to be relatively insensitive to carbon dioxide, whereas fish living in well-aerated conditions have blood which is markedly affected by the CO_2 tension.

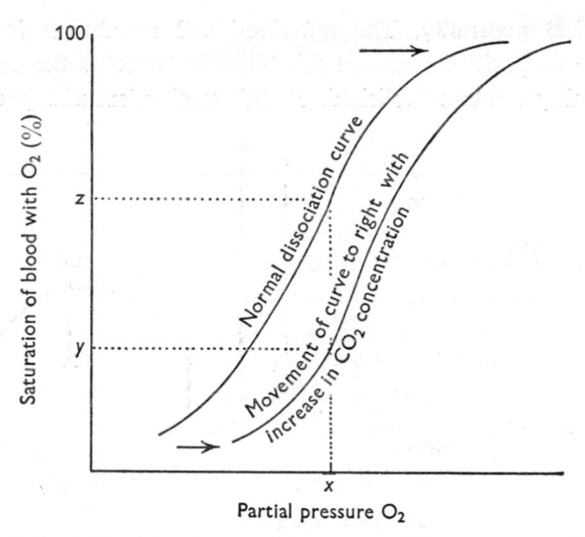

Fig. 27. The Bohr shift. At a given partial pressure of O_2—say x mm Hg—the haemoglobin normally carries $z\%$ O_2, but in regions where the CO_2 concentration is raised, only a smaller amount — $y\%O_2$—can be carried.

3.33 *Transport of carbon dioxide*

Haemoglobin in vertebrate blood also aids the transport of CO_2. It does this partly by combining directly with the carbon dioxide to form a carbamino compound. About one third of the CO_2 that is transported and liberated is in this form:

$$HHbNH_2 + CO_2 \rightleftharpoons HHbNHCOOH.$$

Most, however, is carried in the plasma as bicarbonate. *In the tissues* CO_2 liberated by tissue respiration diffuses into the plasma where it dissolves by a slow process. Some enters the corpuscles and forms carbonic acid by a rapid process catalysed by the enzyme carbonic anhydrase. This subsequently dissociates into bicarbonate and hydrogen ions:

$$CO_2 + H_2O \xrightarrow{\text{Carbonic anhydrase}} H_2CO_3 \longrightarrow HCO_3' + H^+.$$

Oxyhaemoglobin is a stronger acid than reduced haemoglobin and consequently this increase in hydrogen ions encourages the liberation of oxygen by dissociation of the oxyhaemoglobin:

$$H^+ + HbO_2 \rightarrow HHb + O_2.$$

The bicarbonate ions diffuse out of the red corpuscles into the plasma and their place is taken by an inward movement of chloride ions which restore electrochemical neutrality. The red-blood cell membrane is relatively impermeable to positive ions and this 'chloride shift' is the only way in which neutrality can be established. By this mechanism a large amount of

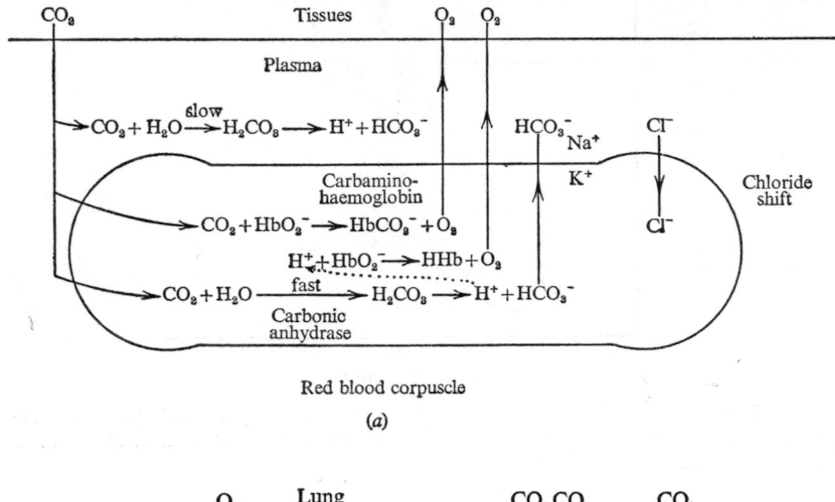

(a)

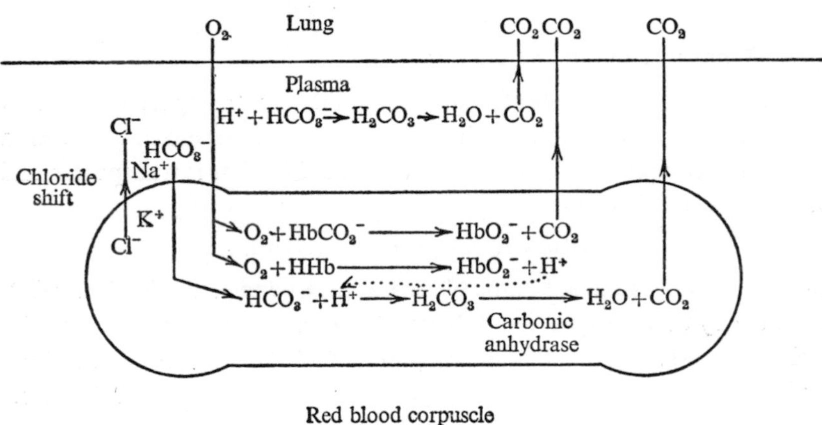

(b)

Fig. 28. The exchanges between (*a*) tissue, plasma and red corpuscles, and (*b*) lung, plasma and red corpuscles in the transport of respiratory gases.

bicarbonate is carried in the plasma where it forms the so-called 'alkali reserve' of the body. *In the lungs*, the proportion of oxyhaemoglobin is increased and being a stronger acid than the reduced form the equilibria of the reactions are displaced in the opposite direction so that CO_2 is liberated from the carbonic acid (Fig. 28). Transport of oxygen and carbon dioxide are therefore complementary to one another.

3.34 *Energy exchanges*

The energy exchanges of the body take place at a cellular level and a discussion of the metabolic pathways involved is found in chapter 11.

The present chapter has dealt with the relationship between the animal and the exchange of its respiratory gases with the environment as well as the transport of these gases in the blood.

By this transport and the uptake of respiratory substrates from the blood the individual cells are provided with the means of tissue respiration and can convert the energy of the food into other forms useful to their own metabolism.

3.4 Respiration in other vertebrates

3.41 *Dogfish*

As we have seen, the respiratory current in a mammal enters and leaves by the same opening and the flow is tidal. In fishes water passes from the mouth to the gill slits and although it enters and leaves these openings intermittently it has been shown that the flow across the gill filaments is almost continuous. In cartilaginous fishes the gill slit in front of the hyoid arch persists as a *spiracle* and water enters through this opening as well as through the mouth. Water which enters the spiracle leaves through gill slits 1–3 on the same side. That entering the mouth generally emerges through the three posterior gill slits (3–5) and again the flow tends to be unilateral. The respiratory movements take place about every one or two seconds. Each cycle includes a rapid expulsion phase followed by a slower period of expansion of the branchial region. The openings between the respiratory chambers and the external medium are protected by valves. Those which cover each of the five gill slits on each side are easily seen and form passive flaps which prevent the entry of water when the gill pouches expand and their pressure is lower than the external water. The spiracular valve is an active valve (i.e. moved by muscular action) which shuts when the mouth cavity decreases in volume. Inside the upper and lower jaws there are small flaps of skin which project backwards; these are not obvious in the common dogfish, but in some species may be very clear, especially among bony fishes.

The wide separation of the gill slits on the outside of the dogfish is because the septum which separates neighbouring gill slits is very well developed. In transverse section (Fig. 29) this septum can be seen to extend from the branchial arch skeleton to the outside where it continues as the gill flap. The septum contains gill rays and muscles which can compress the gill region. On either side of the septum are the *gill filaments.* These are flat plates which are stacked one above the other round each of the five branchial arches. The two rows of gill filaments attached to a given branchial arch form a *holobranch.* The filaments on one side form a *hemibranch* and these come into contact with those of the neighbouring

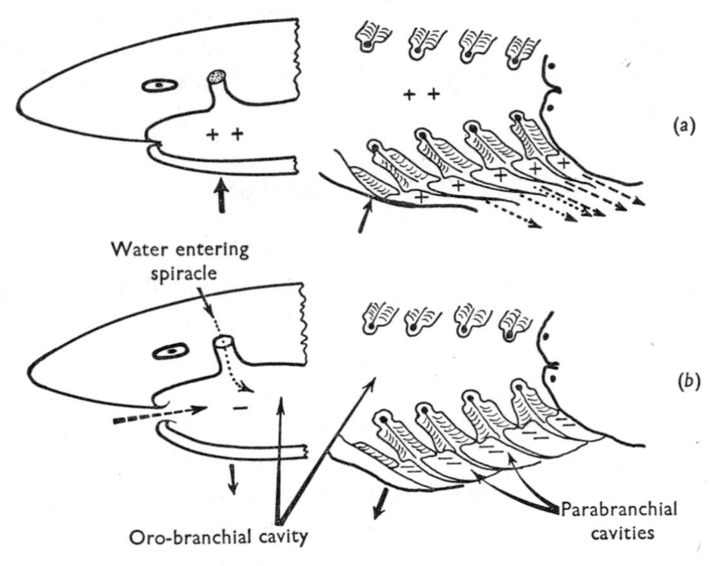

Fig. 29. Ventilation in the dogfish.

branchial arch but only at their tips (Fig. 30). This contact between the filament tips leads to the functional separation of the respiratory cavities into two parts. Water drawn in through the mouth and spiracle first enters the large *oro-branchial cavity* which extends between the gill arches as far as the tips of the filaments. Outside the filaments are found the *parabranchial cavities.* There are therefore five parabranchial cavities on each side which when compressed result in water being forced out through the gill slits. In the common dogfish the slits communicate with only the ventral region of the parabranchial cavities but in other elasmobranchs, for example, the basking shark, they extend along the whole length of the parabranchial cavities. When compressed by hand in an anaesthetised dogfish, water is similarly ejected and the branchial region expands again because of the elasticity of the skeleton. Recordings of the electrical activity in muscles have recently confirmed that the hypobranchial muscles, which run from the pectoral girdle to the jaw and other visceral arches, play little part in the expansion phase of a dogfish during quiet respiration. Activity in the superficial constrictor and most other muscles mainly takes place when the jaws and branchial region are constricted.

Gaseous exchange takes place within the *secondary lamellae*. These fine plates extend upwards and downwards on both sides of the many gill

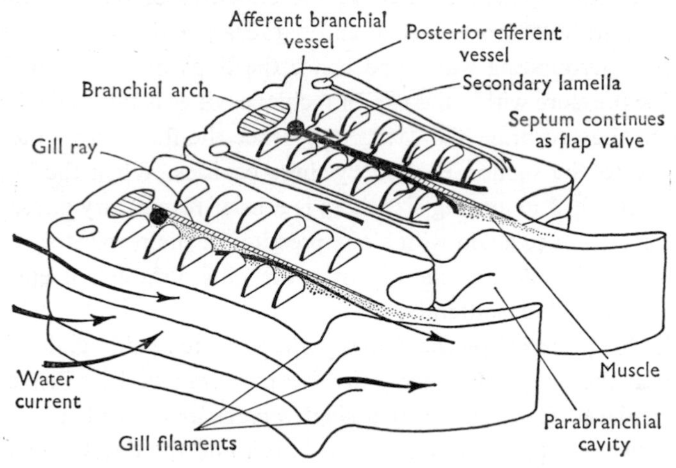

Fig. 30. Two branchial arches and attached gill filaments of a dogfish.

filaments. In the dogfish there are 20 secondary lamellae/mm on each side of a filament. The blood circulates in narrow spaces within the secondary lamellae and comes into close contact with the water. The direction of flow of the blood from an afferent to an efferent branchial vessel is opposite to that of the water and this *counter-flow* enables a greater proportion of the oxygen contained in the water to be removed. The flow of water across the secondary lamellae is almost continuous because of the operation of a double pumping mechanism. The oro-branchial cavity functions as a *pressure pump*, whereas the parabranchial cavities act as *suction pumps*.

Ventilation begins with closing of the mouth and the spiracle. The oro-branchial cavity decreases in volume and water is forced through the gill filaments and between the secondary folds into the parabranchial cavities (Fig. 30). The pressure here is less than in the oro-branchial cavity but as it exceeds that of the outside water the gill flaps open and water is ejected. During this phase of the cycle therefore, the water flow across the gills is mainly due to the pressure-pump action of the oro-branchial cavity. Following this phase the constrictor muscles relax and the whole of the branchial region expands, chiefly because of the elasticity of the visceral arch skeleton and their ligamentous connexions with one another. During this stage the mouth and spiracle open and water enters because of the decreasing pressure within the oro-branchial cavity produced by the lowering of the floor of the mouth and pharynx, together with lateral expansion of the branchial arches. Almost simultaneously the parabranchial cavities enlarge because of the elasticity of the skeleton. The pressure within both oro-branchial and parabranchial cavities is less than that of the surrounding water and hence the flaps over the gill slits are closed. The pressure within the parabranchial cavities is lowered more than that within the oro-branchial cavity and hence the flow across the gills is mainly due to the suction exerted by these cavities. When the branchial region has reached its resting size there is a brief respiratory pause before the cycle begins once again with a compression of the respiratory cavities. Water passes almost continuously over the gills although it enters and leaves the system intermittently.

This double pumping mechanism is perhaps more readily understood by reference to the bony fishes. Here the five parabranchial cavities on each side are replaced functionally by a single *opercular cavity* because of the reduction in the septum between the gill arch and the development of an operculum attached to the hyoid arch. In these fishes there is no spiracle and all the water enters the mouth, which is guarded by maxillary and mandibular valves. About one-fifth of a cycle after the mouth has opened, the operculum expands laterally. Expansion of the opercular cavity lowers its pressure below that of the buccal cavity and water flows across the gill filaments. These filaments are splayed out and are only in contact with one another at their tips. Because of the reduced septum, the water flow between the secondary lamellae is more clearly counter to that of the blood. The efficiency of this mechanism in a trout is indicated by the utilisation of 80 % (50 % in dogfish) of the oxygen contained in the water entering the mouth. If the current is reversed experimentally the utilisation falls by more than half.

The ventilation of the gills in some sharks is produced by their forward movement when swimming with their mouths open and there are no pumping movements, although these take place when the fish comes to rest. In bottom-living cartilaginous fishes such as skates and rays nearly all the water enters through the dorsally situated and very large spiracles. Among bony fishes the mackerel is an example of a form which respires entirely by the current entering the open mouth during swimming. Some bottom-living species tend to have the suction pump more developed; flatfishes are a notable example which also have adaptations preventing the flow of any water from the outside (which might include sand) into the opercular chambers.

3.42 *Frog*

The structure of the lung in amphibians and reptiles is essentially similar to that of mammals. The respiratory surface is much less, however, and in a frog there is only one-fifteenth of the surface area found in the lung of a man for each c.c. of air that they contain. The lungs of some newts are

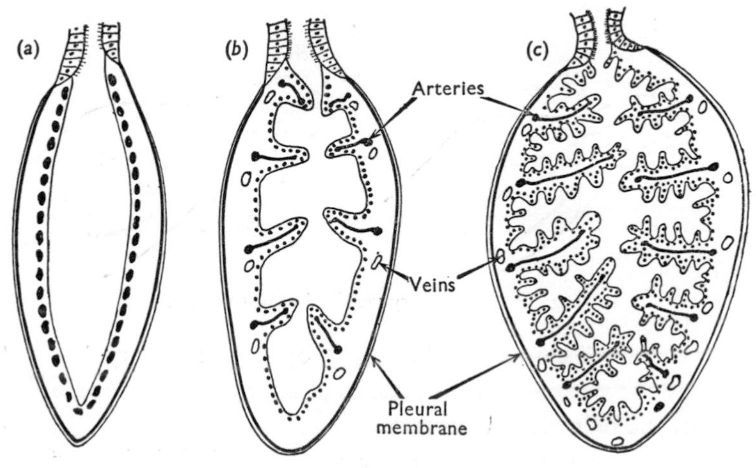

Fig. 31. Stages of increased surface area, by folding, in the lungs of newts and frogs, (*a*) aquatic, (*b*) amphibious, (*c*) terrestrial.

simple sacs and have scarcely any folding of their internal surfaces. The least folding occurs in perennibranchiate forms which utilise their external gills for respiration (Fig. 31). All amphibians make great use of their skins for respiration but this is not so important in reptiles. Reptile lungs have greater surface areas than the frog and true bronchi supported by cartilaginous rings are present. The single trachea leads off from the glottis, unlike the amphibians where both lungs originate directly from the glottis. In both groups, *buccopharyngeal* movements of the throat are present. In frogs ventilation of the buccopharynx by these movements results in a certain amount of gaseous exchange, but in reptiles they are more likely to be concerned with olfaction. The buccopharyngeal movements occur at frequencies between 80 and 120/min in frogs but are periodically interrupted by *pulmonary ventilation.* The mechanism is essentially a buccal force-pump operated by changes in volume of the buccal cavity produced by movements of the hyoid. It is illustrated diagrammatically in Fig. 32 where four main phases are recognised.

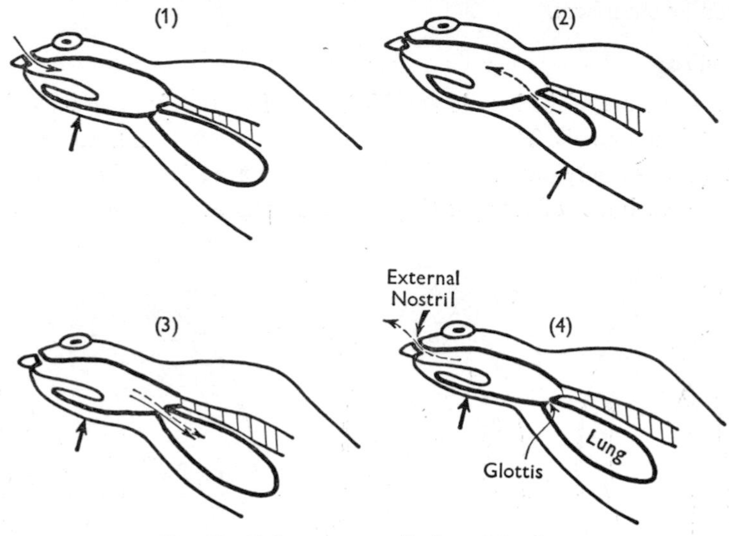

Fig. 32. Pulmonary ventilation of the frog.

(1) The glottis is closed and the nostrils open. The floor of the mouth is lowered by the sterno-hyoideus muscles and air enters the buccopharynx because of the reduction in pressure.

(2) The nostrils are closed, the glottis opens and air is forced from the lungs into the buccopharynx partly because of the elasticity of the lungs and also by contraction of the flanks. Consequently, the floor of the mouth is lowered still further.

(3) The mixed air now contained in the buccopharynx is forced into the lungs through the open glottis. This is brought about by contraction of the petro-hyoideus muscles which raise the hyoid plate. This phase may be repeated several times until most of the oxygen has been removed from the air.

(4) Finally, with the lung filled, the glottis closes and the extra air is forced out through the open nostril.

This pattern is by no means constant and may vary between individuals as well as between different species of frog. Movements of the nostrils play an important part in the mechanism and they result partly from the extra pressure of the lower jaw against the premaxillae, which move the nasal bones so that they occlude the external nostrils. When the hyoid plate is raised and touches the roof of the mouth, the anterior horns block the internal nostrils. These mechanisms, together with the closely fitting joint between the two jaws, ensure that the buccopharynx can be made airtight. Pulmonary ventilation is most important during summer when oxygen is mainly absorbed through the lung. During the winter most oxygen enters through the skin which is the chief route of CO_2 loss throughout the year. (Frogs hibernate encased in mud.)

Table 4. *Pulmonary and skin respiration (c.c./kg/hr) in the frog*

		Cutaneous		Pulmonary	
Date	Weight of frog	O_2	CO_2	O_2	CO_2
Oct.	54 g	54	92	51	15
Apr.	46 g	51	145	160	70

3.43 *Reptiles*

Ventilation of the lungs in reptiles is aided by the action of the ribs which are absent in modern amphibians. These are moved by the intercostal muscles which produce a reduction in pressure within the thoracic cavity resulting in air being drawn into the lung. Relaxation of the inspiratory muscles and activity of abdominal muscles takes place during expiration. In most lizards respiration begins with an initial expiratory phase when air is forced out of the lungs and is then followed by rapid inspiration. The lung is now inflated and remains so during a pause before the next respiratory act. At low temperatures these pauses may be very long but under hot conditions the frequency is quite rapid (30 per min). In the chameleon the lungs have extensions posteriorly into non-vascularised air-sacs which may be inflated by the buccal force-pump (reptiles having no diaphragm). This ability to increase its size is sometimes used as a defence mechanism.

3.44 *Birds*

Air-sacs, which are extensions of the bronchial tree, form an important part of the respiratory system of birds, and occupy as much as 80 % of the total body cavity. The lungs themselves are relatively small and are permanently attached to the dorsal thoracic wall where they are embedded in the ribs. Figure 33 shows the position of the air-sacs and their relationships with the lungs. The detailed mechanism of the bird respiratory system is incompletely understood but it is clear that the air-sacs have a bellows-like action. Air enters when the thoracic cavity is increased in volume when the external intercostals contract, the ribs move outwards and the sternum downwards. The lowered pressure within the thoracico-abdominal cavity draws air into the system which traverses the lungs on its passage to the air-sacs. During expiration air is forced back through the lungs on its way to the trachea. The lungs themselves are made up of a large number of bronchial tubes, the largest being the single *mesobronchus* which is a continuation of the primary bronchus. Arising from the mesobronchus of each lung are two sets of secondary bronchi some of which lead to air-sacs. The anterior and posterior groups

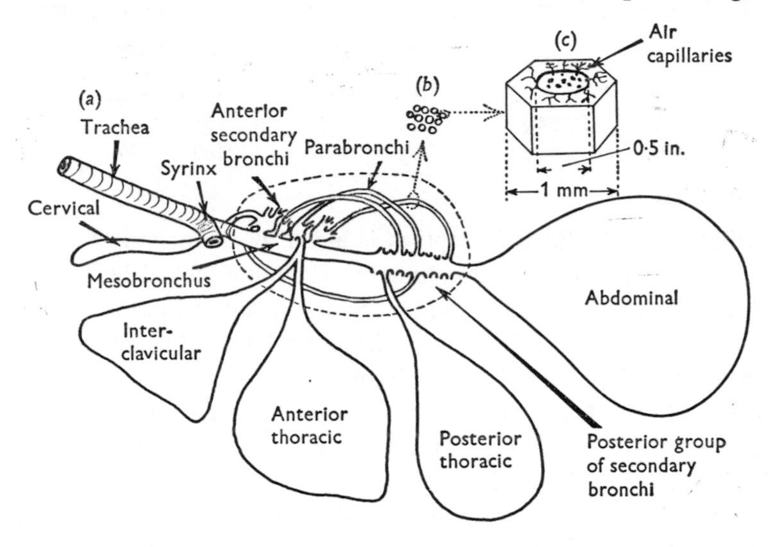

Fig. 33. The main passages of a bird's lung (dashed area) and their connexion with the air-sacs.

of secondary bronchi are joined together in the main body of the lung by a large number (1000) of fine *parabronchi.* Leading off from the parabronchi are fine *air capillaries* which are profusely supplied with blood and are the site of the gaseous exchange. Air is circulated through the parabronchi during both inspiration and expiration so that there are scarcely any dead spaces within this system. Oxygen can diffuse rapidly from the parabronchi into the air capillaries. The distances are less than 0·5 mm and diffusion is sufficient to supply the oxygen required by the animal.

This continuous circulation of air through the parabronchi of birds is reminiscent of the fine adaptations found for ventilation of the gills of fishes.

THE SKIN AND TEMPERATURE CONTROL

The skin of a mammal has a variety of functions to perform; marking as it does the boundary line between the animal and its environment, it controls the way in which the conditions of the latter affect the deeper tissues. Thus the skin protects the body from the entry of micro-organisms; it contains the *sensory receptors* which detect change of temperature, touch and pain; it prevents the passage of water into or out of the tissues and it may be important in temperature control. Besides these functions the skin plays a vital role as a skeletal structure supporting the softer tissues of the body and, in the form of hair, nails and claws, preventing injury or aiding in the capture of food. Finally the skin may be coloured to protect the animal by a cryptic or warning camouflage pattern.

4.1 The structure of mammalian skin

4.11 *The epidermis*

The *epidermis* is the outer layer of skin and this layer can itself be subdivided into different functional and structural units. Directly above the *dermis*, which is mesodermal tissue below the epidermis, is the living *Malpighian layer* and this is composed of cells of cubical shape in which a large number of *mitotic* (normal cell division) figures can be found. These proliferating cells are well supplied with capillary networks from the dermis itself and may also carry the pigment granules responsible for the colour of the skin. Above the Malpighian layer and derived from it, are the cells of the *stratum granulosum* which are characterised by the presence of granules and contain a nucleus and well-marked boundary. The next layer is the *stratum lucidum* and here the presence of the protein *kerato-hyalin* gives the layer its clear appearance, while the cell shapes have become irregular and no nuclei are visible. The upper layer is the *stratum corneum*, which contains the important substance *keratin*. Cells of this layer are *squamous*, that is, flattened, and with individual structure completely lost.

In mammals the development of a lightweight waterproof layer in the outer epidermis has been of great importance in successful colonisation of terrestrial environments. Keratin is quite impermeable to water and gases and its presence insulates the animal from water exchanges with its environment via the skin. While it is true that reptilian scales have the same effect, they are very much heavier than the mammalian epidermis. The actual depth of the epidermis varies in the body and there is an adaptive increase in thickness over the soles of the feet and joints which protects against the wearing effects of the environment.

Variations in the epidermis from the type described above are found in two major forms: first the hairs, diagnostic of mammals, and then the thickening of the keratinised layer as claws, nails and hooves.

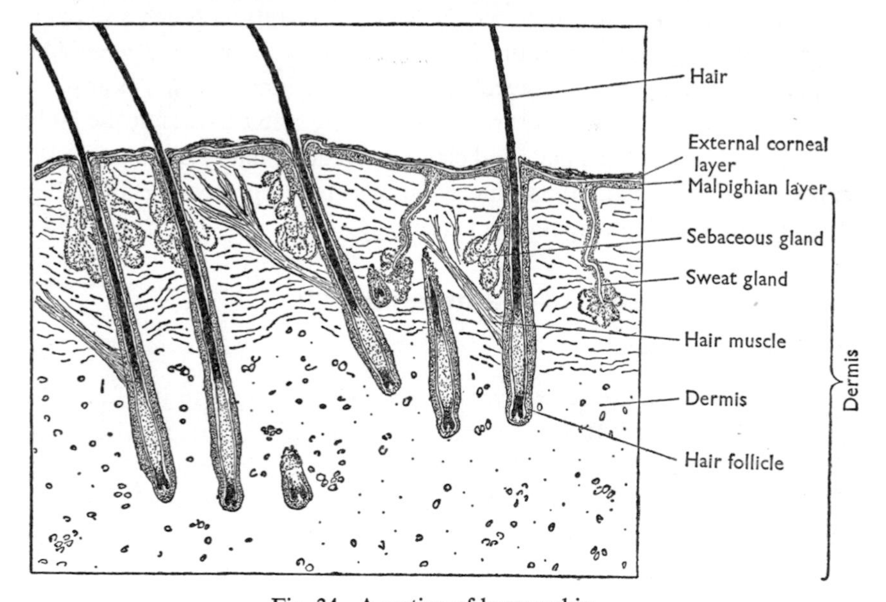

Fig. 34. A section of human skin.

4.111 *The hairs.* These are epidermal in origin, originating from invaginations of the Malpighian layer in deep pits or *follicles.* At the base of these follicles a mass of Malpighian cells proliferates, being well supplied with a plexus of blood capillaries, and elongated keratinised cells are budded off. The central cells of the hair contain air and the outer ones are usually pigmented, so that by inherited arrangements of appropriately coloured areas camouflage patterns are formed (Fig. 34).

In general the shaft of the hair points backwards, minimising friction. Different species have their own particular hair patterns and it is interesting that the hair pattern of man is very similar to that of other closely related apes (Fig. 34).

By the attachment of sensory cells to the base of the follicle the hair acts as an extended touch-receptor and this capacity is particularly well developed in the *vibrissae*, e.g. on a rat's snout. The main function of hairs, however, is to provide a thick insulating layer over the body which cuts down the loss of heat to the environment. Attached to the follicle is a small muscle, the *erector pili*, by which the individual hair can be erected, which will vary the efficiency of the whole insulation. These hair muscles are under sympathetic control and not only contract to make the hairs retain a thicker layer of warm air next to the skin, but also cause them to be raised as an indication of aggression. 'Gooseflesh' seen in man is due to the contraction of hair muscles when cold and is a vestigial physiological response no longer serving in insulation. (A full treatment of the control of temperature is found on p. 78.)

Associated with the hairs are *sebaceous glands* which lie some way up the shaft of the follicles. The *sebum* they secrete is an oily substance and makes the hairs more flexible and waterproof. The *mammary glands* are thought to be evolved from sebaceous glands.

4.112 *Claws.* The claw, also found in birds and reptiles, is the basic epidermal modification from which the nails and hooves of mammals have been derived. Unlike nails the claw covers the whole end of the digit and consists of a germinal layer which proliferates a keratinised layer which continuously moves outwards towards the tip. On the underside of the claw a soft tissue represents the transition from claw to skin. Claws are important for defence of the body or for catching prey and are also used in climbing and grooming.

4.113 *Nails.* These are formed from a very thick stratum corneum, especially its deeper regions, and consequently have something of the characteristics of the stratum lucidum below, for example, the remains of nuclei are found. No stratum granulosum is present so the nail lies directly on the Malpighian layer below. Nails are found at the end of digits but only on the upper surface, assisting the animal to grip objects and in grooming. The nail is a special type of claw.

4.114 *Hooves.* Like the nail, hooves are adaptations peculiar to mammals, especially the ungulates which walk on the ends of one or two lengthened digits. Hooves are much broader and less pointed than claws and are hemispherical with a hard keratinised layer surrounding a soft transition tissue in the middle. The latter may act as a shock absorber.

4.115 *Horns.* The horn is a thickening of the epidermis and the outer keratinised layer surrounds a bony outgrowth of the head. Horns are very strong though much lighter than bone and are used for protection and for fighting. (The horn of the rhinoceros is made of tightly packed hair.)

4.116 *Other functions of the epidermis.* Besides the roles of the epidermis outlined above it also plays a part in the defence of the body against micro-organisms. Keratin is not easily digested by these organisms and they do not find it possible to establish themselves under normal conditions. The secretions of sebaceous and sweat glands are inimical to their growth.

4.12 The dermis

This thick region below the epidermis is made largely of *connective tissue*, that is, it has the skeletal properties associated with *collagen* and *elastin* proteins. The scraping off of the hairs and subsequent tanning of the skin by chemicals make cross-linkages between these proteins, converting the skin into leather, an exceedingly tough biological substance. Even without tanning the dermis is very strong and supports the underlying tissues to maintain the characteristic shape of the animal.

Rich networks of capillaries permeate the dermis from vessels in the deeper tissues and by the contraction or dilation of these through the action of sympathetic nerve fibres the heat losses that take place directly from blood to environment can be controlled. The lower parts of the dermis form subcutaneous stores of fat and help to insulate the body in man but, on the whole, this adaptation is more typical of marine mammals.

Important structures of the dermis are the *sweat glands*, although much variation exists in the number and position of these among the mammals. The sweat gland consists of a secretory coiled tube fed by a capillary network and supplied with individual nerve fibres, the tube leading to the surface by a long sweat duct. Despite the fact that these glands are embedded in the dermis they actually derive from the Malpighian layer. The scents they emit may be important in the reproductive behaviour of the animal.

Various sorts of sensory cells are found in the dermis and the skin is an obvious site for receptors of touch, temperature and pain. Tactile corpuscles may be oval (the *Pacinian bodies*), while other receptors with different shapes respond to cold (*Krause's organs*) or heat (*organs of Ruffini*). There is a dendritic plexus associated with pain reception but this also results from the overstimulation of any type of receptor.

4.2 Modifications of the skin in vertebrates other than mammals

4.21 *Elasmobranchs*

These have characteristic *placoid scales* which are constructed on the same basis as a tooth and afford good protection to the animal. Certain cells from the dermis called *odontoblasts* form a pulp cavity and secrete a thick layer of dentine permeated by protoplasmic canals and surrounded

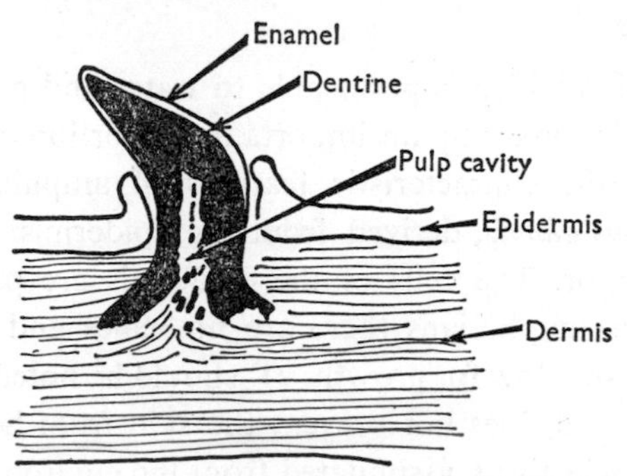

Fig. 35. Elasmobranch skin: placoid scale.

by enamel. The scales originate at the junction of the dermis and epidermis and the latter is itself a thick keratinised layer which gives added protection. In the region of the jaws the dermal region produces along its edge transitional types of structure between teeth and scales, while further in long teeth are formed. The tooth is thus clearly seen to be homologous with the placoid scale.

4.22 *Teleosts*

These have *cycloid scales* and these consist of flat 'bony' plates in the dermis which are covered by a keratinised epidermis. As the scales are only rooted at one end they allow flexibility of movement but, unlike the placoid scale, they do not penetrate the epidermis. In this cycloid scale, which is found in most living bony fishes, there is an outer calcified layer over a disc of fibrous connective tissue. Seasonal changes in growth rate are shown on the scale by a series of rings so that the individual scale can be used to determine the age of the fish. In both elasmobranchs and teleosts the skin is impermeable to the entry of water.

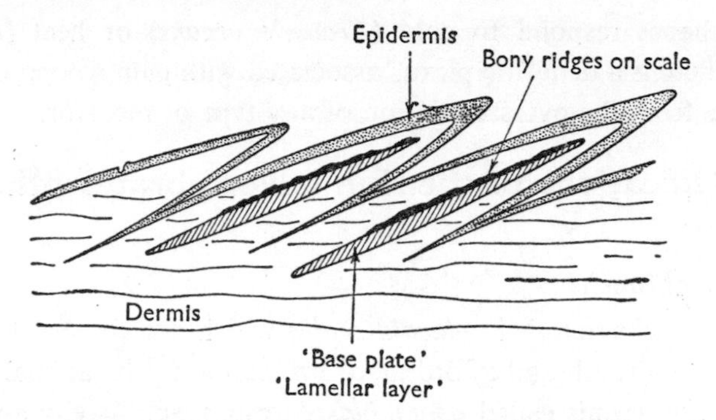

Fig. 36. Section of teleost skin: cycloid scales.

4.23 *Amphibians*

These have a skin which is permeable to water and gases and in these vertebrates the skin makes up an important proportion of the respiratory surface. One of the characteristic features of amphibian skin is the presence of *mucous glands*, derived from the epidermis, and opening by ducts to the exterior. The cells of the epidermis are partly keratinised though in modern amphibians there are no scales and periodically the skin must be shed to allow for growth. (It should be noted that the keratin layer is very thin and does not prevent the skin from being permeable.) Mineral salts are selectively assimilated from the environment.

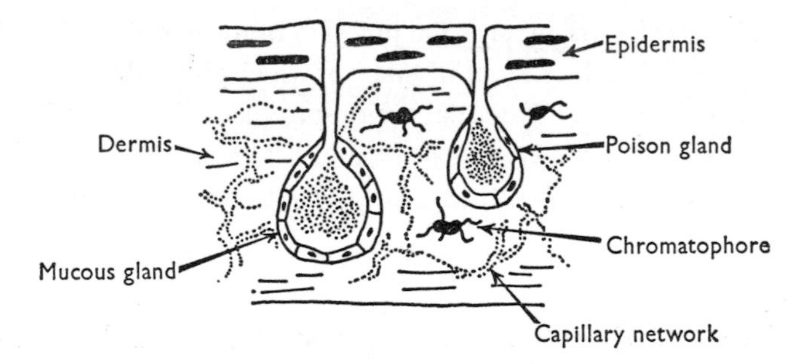

Fig. 37. Section of amphibian skin.

The other type of gland found in amphibian skins is poison-secreting; these organs may be concentrated in definite regions, for example, along the back in many species of Rana. In some species the mucus itself is poisonous. Such poisons are of benefit to their owners in preventing attack by predators and their presence is associated with bright warning colorations.

Amphibian skin may also be modified into horny tubercles or thickened to provide *nuptial pads*. It may contain scent glands. Finally, it may contain in its dermis various *chromatophores* which enable its owner to produce camouflage or other colorations.

4.24 *Reptiles*

These have scales which, unlike those of fishes, are formed mainly from epidermal layers. The scales have an outer dead horny part made largely of keratin and are periodically shed to allow for growth. In some reptiles, such as the snakes, the whole epidermis is sloughed off at one time as with the amphibians and paralleling the moulting of arthropods. Reptile skins are almost impermeable to water, which has allowed their efficient colonisation of the land. Although their skins do contain glands of various types no sweat glands are present and the skin does not provide a means of temperature control. In some cases the scales are enlarged to form protective spines and horns while in others addition of dermal elements gives greater reinforcement, for example, crocodiles and turtles. As with amphibians chromatophores may be present and enable colour changes to be made—for example, in the chameleon.

4.25 *Birds*

Birds have *feathers*, which are diagnostic of the class and which have arisen from the scales of the reptiles from which the birds evolved. The feather arises as an outpocket of the skin called a *papilla* which contains an inner Malpighian layer and outer periderm and stratum corneum. Within the Malpighian layer is a projection of the dermis with a good supply of blood capillaries and it is this region that pushes out projections of the surrounding epidermis. The stratum corneum thus becomes arranged into a long *rachis* with extending *barbs*. From these barbs further projections called *barbules* arise which themselves carry small hooks or *barbicels*. The latter allow attachment between members giving a light airfoil, the *vane* whose efficiency is maintained by the preening activity of the bird. An oily substance from the *uropygial* gland near the anus is used in preening and helps to keep the vane flexible.

After the feather structure has been laid down the internal dermal region of the Malpighian layer withdraws and forms a sunken follicle below the surface of the skin. As the blood vessels and nerves are associated with this layer the distal part of the feather is no longer living following its withdrawal.

Variation on the flight feather described are found in the down feathers, which are mainly for insulation.

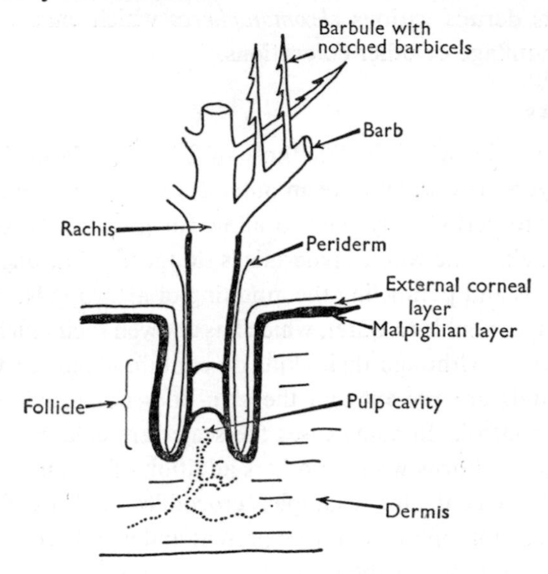

Fig. 38. Section of bird skin and plume feather (see also Fig. 97).

4.3 Control of temperature

One feature which characterises mammals and birds is that they are warm-blooded or *homoiothermic*. This refers to the fact that their body temperature is maintained nearly constant despite wide fluctuations in the temperature of the external environment. The lower vertebrates and the invertebrates are *poikilothermic*, as their body temperatures depend much more on the external temperature. In the graph of Fig. 39, the body temperature for a lizard is seen to vary directly with the external temperature. Poikilotherms usually have body temperatures within a few degrees of the environment; but sometimes they may be quite different, especially when the external temperature fluctuates more rapidly and the animal is relatively large. Smaller individuals follow the air temperature with less time-lag than larger forms, which tend to maintain a temperature which is the mean of any fluctuations in the environment. One effect of this is that small poikilotherms are able to take advantage of brief periods of sunlight to raise their body temperatures. Many reptiles, for example, are very active early in the morning when there is a lot of sunlight and their body temperatures are raised, but as the environment becomes too hot in the afternoons they may retreat into burrows and emerge again in the evening as the environmental temperature falls. Finally, by remaining under cover during the night they avoid excessively low temperatures.

The metabolic rate of poikilotherms is directly related to the temperature and their activity is reduced at temperatures below that of their normal

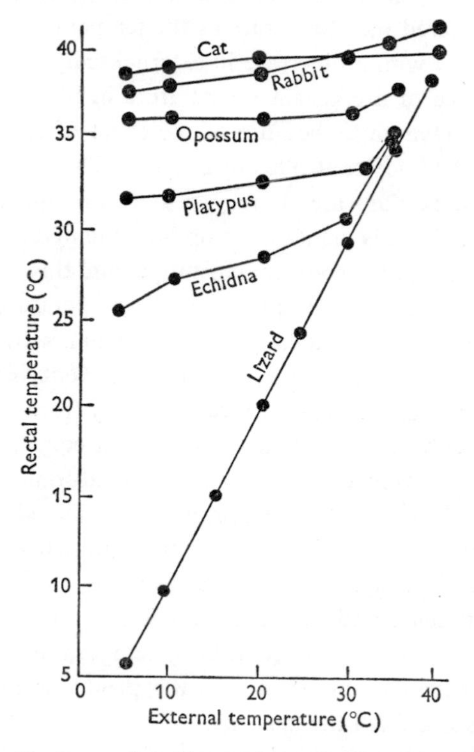

Fig. 39. External and internal temperature relationship in various vertebrates.

environment. Considered over a longer term, however, poikilotherms do have a system of regulation. This is because animals which have lived for some time in such cold conditions have metabolic rates which are higher than those of closely related species, or individuals of the same species, which are suddenly subjected to a lowered temperature. This phenomenon, which is called *acclimation*, means that poikilotherms are able to extend their geographical range over and above the temperatures in which they are normally active. Cold-adapted animals cannot withstand such high temperatures as warm-adapted ones and vice versa.

Warm-blooded animals on the other hand tend to maintain a constant metabolic rate in a zone of so-called thermal neutrality (23–27° C in man) when the metabolic rate of a resting and unclothed subject is least. On either side of this range, however, the oxygen consumption rises as the animal makes compensatory responses which tend to maintain the constant body temperature. This is distinct from the situation in poikilotherms, which have a lowered metabolic rate as the temperature falls and an increased metabolism with an increase in external temperature.

The maintenance of a constant temperature in birds and mammals is thought to have advantages because of the effect of temperature on the reactions within the body controlled by enzymes. These are proteins and at higher temperatures they are destroyed and consequently the rate of reaction declines. There is therefore an optimum temperature at which the rate of reactions within the body are most rapid and this is usually between 30° and 40° C. This is also the range of body temperature found in birds and mammals and there is evidently a delicate balance between the maximum efficiency and the rate of breakdown of these enzymes. The range of temperatures which animals can survive is roughly from 0 to 50° C, which again coincides with the range of activity of most enzymes.

The body temperature of a mammal is not absolutely constant; for instance, in man 98·4° and 99·6° F are often considered the normal oral and rectal temperatures, but in fact they may range from 98° to 99° and 99° to 100° F respectively. The temperature varies throughout the 24 hr, being lowest between 6 and 7 a.m. and highest about 12 hr later. Maintenance of this more or less constant body temperature results from the interaction of processes concerned with the production of heat and the prevention of its loss from the mammal.

Most heat is lost from mammals by radiation and conduction from the body surface, relatively small amounts being lost by evaporation. The relative amounts lost by these three processes change with temperature, however, mainly with the result that the amount lost by evaporation exceeds that by radiation above 31° C. In cooler air a man may lose 1 litre of water per day by evaporation but during hard work in a desert his loss may be as great as 1·6 litres per hour.

The loss of heat from the body is limited by the insulation afforded by the fur or feathers. Heat is produced by the metabolic activity of the animal, which increases as the external temperature is lowered. Its maximum amount is produced by *asynchronous* activity of muscles which eventually results, reflexly, in *shivering*. In animals which live habitually

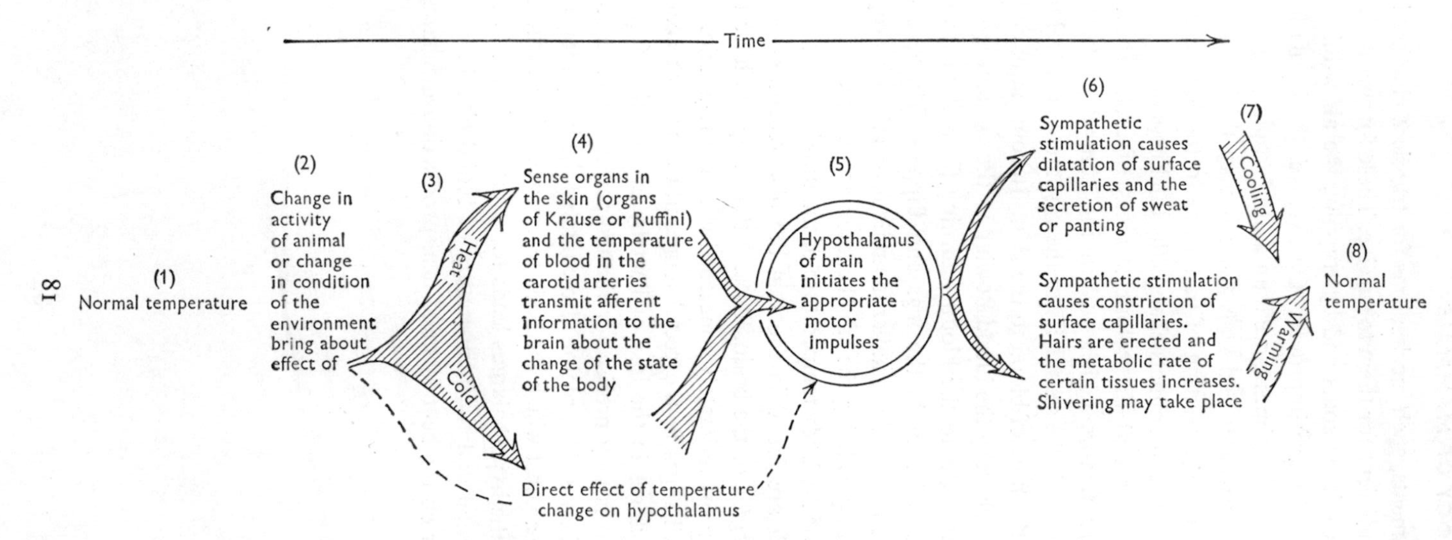
Time
(1)
Normal temperature
(2)
Change in activity of animal or change in condition of the environment bring about effect of
(3)
Heat
Cold
(4)
Sense organs in the skin (organs of Krause or Ruffini) and the temperature of blood in the carotid arteries transmit afferent information to the brain about the change of the state of the body
Direct effect of temperature change on hypothalamus
(5)
Hypothalamus of brain initiates the appropriate motor impulses
(6)
Sympathetic stimulation causes dilatation of surface capillaries and the secretion of sweat or panting
Sympathetic stimulation causes constriction of surface capillaries. Hairs are erected and the metabolic rate of certain tissues increases. Shivering may take place
(7)
Cooling
Warming
(8)
Normal temperature

Fig. 40. The homeostatic response of the body to change in temperature.

in cold environments, such mechanisms for transient changes in temperature fall are insufficient for them to survive. These forms usually are larger and have much thicker coats and by trapping the air within the coat its insulating properties are improved. On average, the insulation of arctic mammals is about nine times more than that of tropical forms. In aquatic mammals such as whales and seals there is very thick *subcutaneous fat*, or *blubber*, because the wetting of hair leads to the loss of its insulating properties. Birds and mammals change the thickness of their coats during summer and winter and the colour changes of arctic species are well known. Further factors which govern the control of heat loss from mammals and birds is the sensitivity of the vasomotor control mechanisms which regulate the flow of blood to the skin. In some aquatic forms there is a mechanism whereby the blood flowing to the extremities transfers the major part of its heat to the blood returning from the extremity and so reduces the loss of heat in the legs and flippers. The size of external structures of the mammal is smaller in arctic forms than in tropical forms, well-known examples being the small ears in arctic hares.

The control of the body temperature of mammals results from sensitive receptors which are found in the skin but also receptors present in the *hypothalamic region* of the brain. These latter receptors have been shown to be the ones most intimately concerned with the regulation of the body temperature. Changes in the rectal temperature are not accurately correlated with changes in the loss of heat by evaporation from the body. This correlation is very precise for the temperature measured very close to the hypothalamus. This region also contains the body thermostat which is the reference against which the body temperature is measured. Damage or removal of the hypothalamus leads to variations in the body temperature of a mammal and the temperature of blood circulating to it has a profound effect on the heat loss and heat-production mechanisms.

5 THE BLOOD AND CIRCULATORY SYSTEMS

5.1 The heart and circulation of vertebrates

5.11 *The double circulation of mammals and birds*

The speed of circulation is highest in mammals; in man the blood takes about 18 sec to complete its circulation during maximum exercise, but 86 sec when at rest. During this time it passes successively through the right ventricle, lung capillaries, left auricle, left ventricle, body capillaries, right auricle and back to the right ventricle (Fig. 41). This type of circulation, in which the blood passes twice through the heart during one complete cycle, is called a *double circulation* and was discovered by William Harvey in 1628. A rapid circulation is ensured because the pressure and velocity of flow is raised by the heart after the blood has passed through the fine capillaries in the respiratory organ. Circulation may therefore be divided into two parts—the *systemic circuit* and the *pulmonary circuit*. One essential condition for such a circulation is that equal volumes can pass round each circuit in unit time and this is only satisfied by the cardiovascular systems of adult mammals and birds.

5.111 *Chambers of the heart.* The hearts of mammals and birds may be considered as two parts (the right and left heart) which act in series with one another although they are bound together as a single organ. Each pump is made up of a relatively thin-walled auricle to which blood is returned from the capillary systems and a thick-walled muscular ventricle which is responsible for raising the pressure in each circuit. The ventricles contract almost simultaneously and blood is forced into the aorta and pulmonary artery from the left and right sides respectively. During ventricular *systole* (i.e. contraction), blood cannot return to the auricles because of the presence of *auriculo-ventricular valves*. These are formed of flaps of tissue which allow the blood to pass from the auricle to the ventricle but close these openings when the pressure in the ventricle is greater than that in the auricle. Eversion of the valves into the auricle is prevented by fibrous strands (*chordae tendinae*) attached between the valves and the *papillary muscles*. The valve on the right side is called the *tricuspid* valve in man and that on the left side (*bicuspid* or *mitral* valve) is held in position by even larger papillary muscles. This is because the left ventricle is not only the largest chamber of the mammalian heart but also the most muscular. The left ventricle is also very large in the bird heart but one interesting difference is that the right auriculo-ventricular

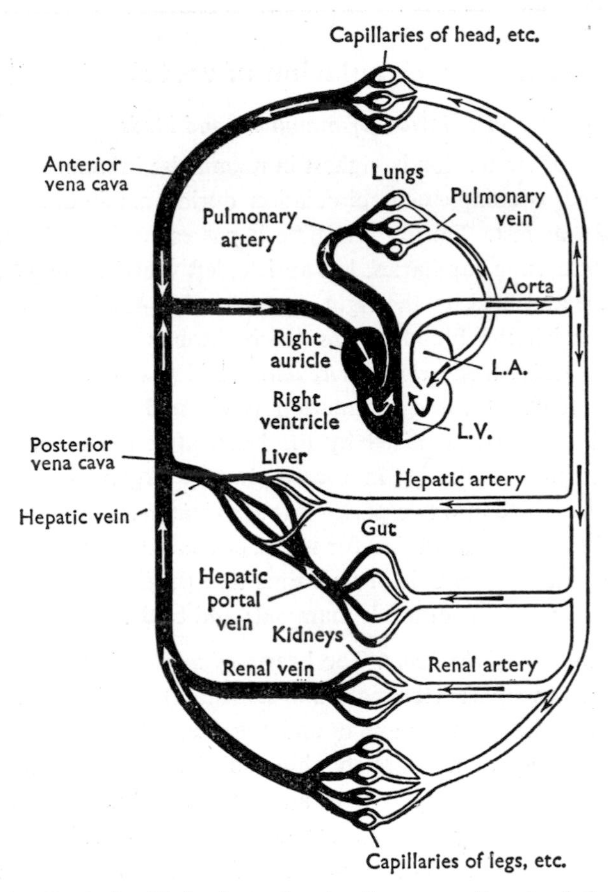

Fig. 41. A simplified scheme showing the mammalian circulation.

valve is a muscular ridge which almost entirely circles the orifice and functions like a sphincter. Consequently there are no chordae tendinae on the right side but the left side is the same as a mammal heart. The exit from the ventricles is protected by a series of pocket or semi-lunar valves which prevent the return of blood to the ventricles on the completion of systole.

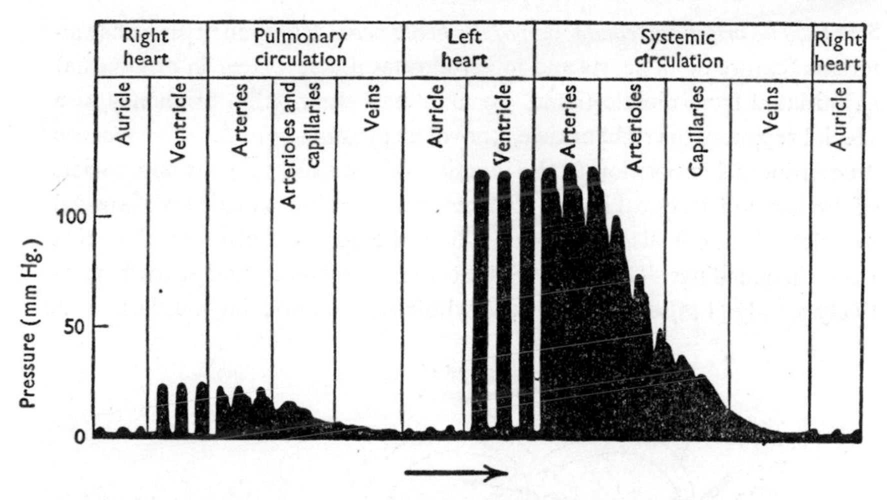

Fig. 42. Diagram to illustrate pressures in different parts of mammalian circulation. Pressure fluctuations produced by three heart beats are shown in each section. Note the fall in pressure as blood traverses the arterioles of the systemic circulation.

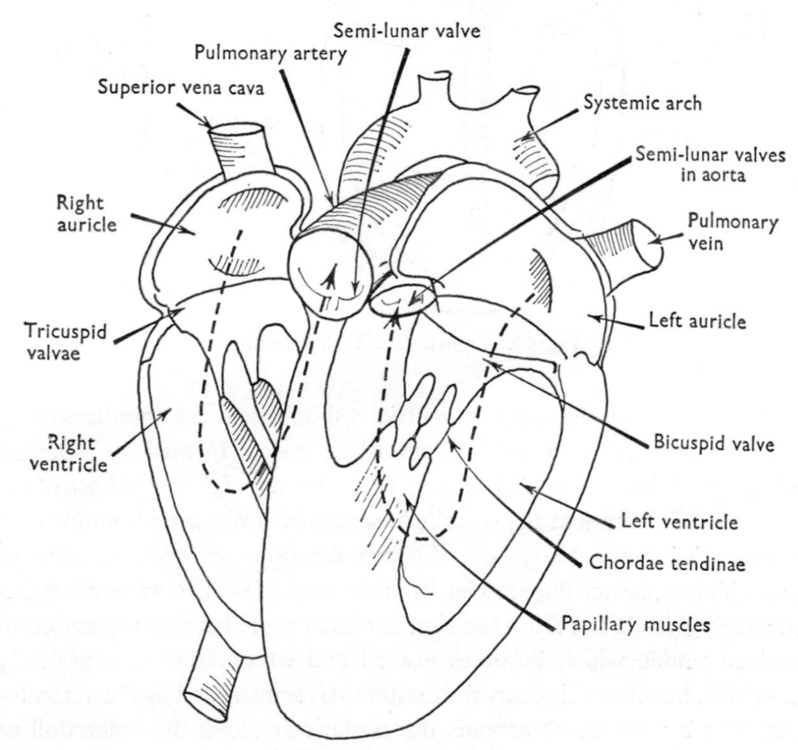

Fig. 43. Rabbit heart.

5.112 *The origin and conduction of the heart beat.* Rhythmicity is a characteristic feature of all hearts and in vertebrates it is *myogenic* in origin, that is, initiated from muscle tissue. In mammals the rhythm originates in a special region of the right auricle, known as the *sinu-auricular node* because it corresponds in position to the junction of the sinus venosus and auricle of the fish and ancestral heart. This *pacemaker* region is made up of special muscle cells and is also the region where the nerves controlling the heart have their endings. The beat originates spontaneously and spreads relatively slowly (1 m/sec) through the whole of the auricular muscles. It is

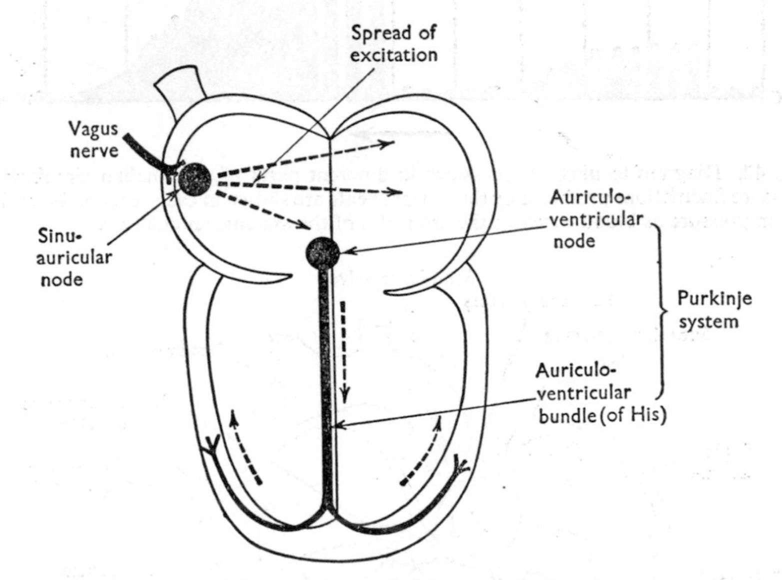

Fig. 44. Control of the heart beat.

especially important that the two ventricles should contract simultaneously and this is ensured by a special conducting system formed by modified cardiac muscle fibres (*Purkinje tissue*). This takes its origin at the base of the inter-auricular septum at the so-called *auriculo-ventricular node* and is vital for transmitting the heart beat through the non-conductile connective tissue which separates the auricles from the ventricles. The wave of excitation which spreads out from the sinu-auricular node reaches the auriculo-ventricular node which becomes excited and conducts relatively slowly (0·2 m/sec), but after this delay the excitation is transmitted by the auriculo-ventricular bundle which spreads out posteriorly along the inner wall of both sides of the inter-ventricular septum. The wave of excitation is conducted at 5 m/sec along these fibres and in the absence of any sinu-auricular pacemaker they may develop their own beat spontaneously. The passage of the wave of contraction follows the electrical events after a shorter delay than in skeletal muscle. The electrical phenomena can be recorded from electrodes placed at different points on the body surface as the electrocardiogram (e.c.g.) and give very valuable information concerning the state of the heart. It appears that there are three phases in the spread of ventricular activity—first, from right to left in the inter-ventricular septum, then from the inside to the outside of the ventricular walls and finally from the tip (i.e. posterior) of the ventricle towards its base, as this region is not supplied by conductile tissue. As a result of this activity both ventricles contract almost simultaneously, beginning first of all with the inner layers of the more posterior regions of the ventricles. During ventricular systole the papillary muscles also contract and maintain the position of the auriculo-ventricular valve. All of these adaptations ensure a more powerful beat of the heart.

5.113 *Pressures within the circulatory system.* The aortic pressure produced during systole is about 120 mm Hg falling to 80 mm Hg during *diastole* (i.e. relaxation of the heart). The pressure difference between these two during each contraction is the *pulse pressure.* Pressures in the pulmonary circuit are much lower, systolic 27 mm Hg, diastolic 10 mm Hg, because the pulmonary circulation is much more distensible than the systemic (see Fig. 42). During its passage through the systemic circulation the pulse, as well as the systolic and diastolic pressures, decreases progressively. The smoothing out of the pulse is due to the elasticity of the arterial walls and the general fall in pressure is not great in the arteries because of their relatively large diameter. In the small arterioles and capillaries the greater resistance of the fine tubes reduces the pressure because of the increased friction at their walls (Fig. 42). By the time blood reaches the veins the pressure is extremely low (less than 10 mm Hg) and its return to the heart is aided by other features of the circulation. These are absolutely vital because within the large veins the pressure may be as low as 5 mm water. The tone of the body muscles and especially the increase in muscular contraction within the extremities during exercise play a very important role. Within the veins there are valves preventing the backflow of blood and hence any compression causes flow towards the heart. Another feature aiding the venous return is the lowered pressure within the thoracic cavity which results from the lung elasticity as described previously. The negative pressure of 6 mm Hg falls to 2·5 mm during expiration but during both phases it will draw blood towards the heart.

5.114 *Heart output and its regulation.* Essentially, then, a circulatory system consists of a pump and a series of tubes varying in diameter, which convey the blood to the tissue capillaries and then back to the pump again. The passage of blood from the pump is intermittent but its movement through the capillaries is continuous. The elasticity of the thick-walled arteries plays a vital part in the maintenance of this continuous supply of blood to the tissues. During systole, energy is stored and can be released during diastole as the blood is squeezed through the arterioles. Clearly the volume of blood passing to the tissues in a given time depends on the amount pumped by the heart and the resistance to its flow through the capillary system. The cardiac output may be varied either by changes in frequency or output per beat. Frequency varies a great deal among different mammals; in general the larger the animal the slower its heart rate. In small shrews and in birds the frequency may be between 800 and 1000 beats/min. For a given animal there appears to be a limit to which the frequency can be raised and in man it is about 190–200/min. At rest, the output of a man's heart is about 5 litres/min and it may be increased as much as 7 or 8 times during severe exercise. In trained subjects this maximum output can be maintained at lower frequencies of heart beat. It functions more efficiently at these frequencies, partly because there is greater time for recovery between each beat. Many athletes at rest have pulses less than 50 per min, the lowest recorded being about 39, as compared with the normal frequency of 70 per min. The output per stroke is greater in athletes and those accustomed to strenuous work and in some cases is associated with hypertrophy (enlargement) of the heart.

The arterial blood pressure is monitored by receptors in the carotid sinus and aorta which respectively send sensory fibres to the medulla in the glossopharyngeal nerve (IX) and depressor branch of the vagus (X). Excitation of this pathway leads to a reduced cardiac output and consequent fall in the arterial pressure. Other important reflex mechanisms include those which ensure that the heart becomes accelerated when the large veins and auricles are distended. An intrinsic regulation is also produced by the so-called law of the heart whereby the tension exerted by the individual muscle fibres increases at greater lengths. Changes in heart frequency mainly arise by variations in the so-called *vagal tone.* The endings of the vagus nerve near the sinu-auricular node secrete acetyl choline (originally called 'vagusstoff') which has an inhibitory effect upon the pacemaker region. Most of the effects of afferent stimulation result from an increase or decrease in this tone. These vagal fibres form part of the parasympathetic system and the reflexes in which they are involved are fairly specific. The sympathetic innervation has an accelerating effect due to the local liberation of adrenaline but it is often associated with liberation of adrenaline from the adrenal medulla as part of a general response changing the condition of the body in an emergency. Not only does it produce an increase in heart rate but also helps the speed of blood flowing to the muscles by causing the plain muscles of the arteriola walls to relax (*vasodilatation*). Altering this peripheral resistance of the vascular system is an important way in which the blood flow is regulated. Not only the total output of the heart, but also its precise distribution can be regulated by varying the diameter of the different capillary systems.

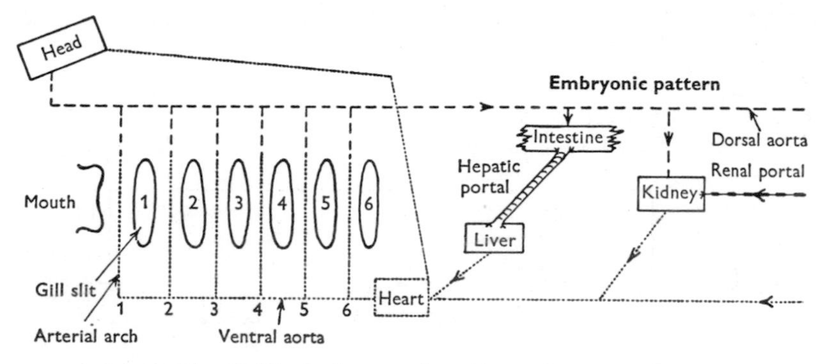

Fig. 45. The basic type of vertebrate circulation.

5.12 *Development of the aortic arches*

5.121 *Lower vertebrates.* The double circulation of mammals and birds is very distinct from the single circulation of fishes and in the absence of other evidence it would be difficult to envisage how one could evolve from the other. Fortunately, however, many intermediate stages are available among living vertebrates both as adults and in their developmental stages. A study of the latter shows that in all vertebrates six aortic arches join the ventral aorta with the lateral dorsal aorta on each side (Figs. 45 and 49). In fishes these embryonic arches become broken up into the capillary networks of the gills. In adult tetrapods those which persist are continuous between the ventral and lateral aorta. Usually these represent the third, fourth and sixth embryonic arches and are known respectively as the *carotid*, *systemic* and *pulmonary* arches in adult forms. The carotid supplies the head, the systemics take blood to the rest of the body, and the pulmonary arteries go to the lung. The pulmonary artery may retain its connexion with the lateral dorsal aorta. This connexion, found during development and also in some adult amphibians, is called the *ductus arteriosus* or ductus Botalli (Fig. 46). Blood returning from the lungs enters the left auricle in all lung-breathing forms. In some amphibians the fifth aortic arch persists (e.g. sala-

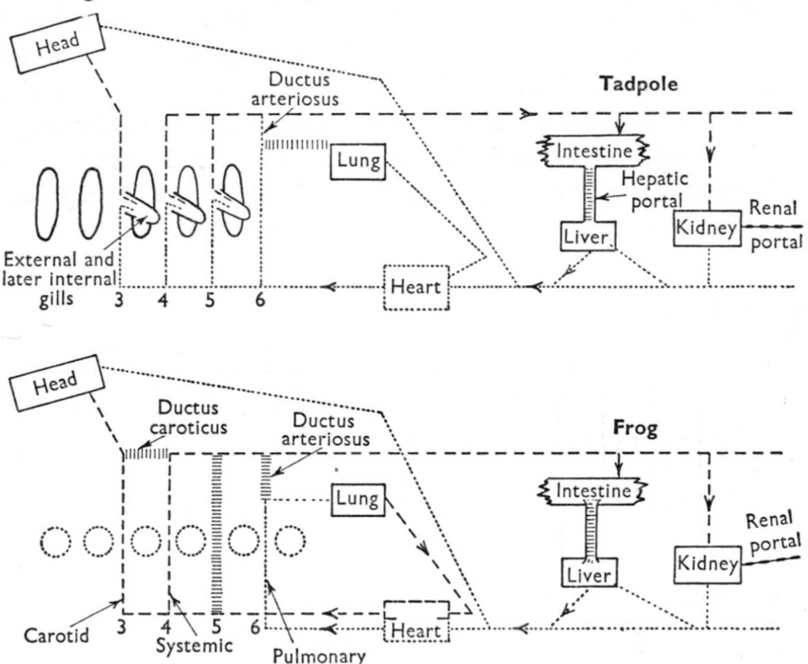

Fig. 46. The aortic arches of the tadpole and frog, showing how they can be derived from the basic 'fish' condition. ||||| = vestigial tissue.

manders and newts) and in some reptiles (e.g. lizards) the connexion between the systemic and carotid arches remains as the *ductus caroticus*.

During development, tadpoles pass through a stage in which the arches form gill capillaries and may also supply the external gills. Later the gill circulations are lost and the adult pattern develops. In mammals the six homologous arches are found but are never all present at the same time.

5.122 *The foetal circulation.* There are two remarkable features of the circulation of the mammalian foetus associated with the way it receives oxygen. The lung is non-functional and therefore little blood can pass round the pulmonary circuit. Secondly, blood oxygenated at the placenta passes directly into the inferior vena cava. The way in which these features are combined in the foetal circulation and the changes which occur at birth provide excellent examples of the adaptation of the circulatory system

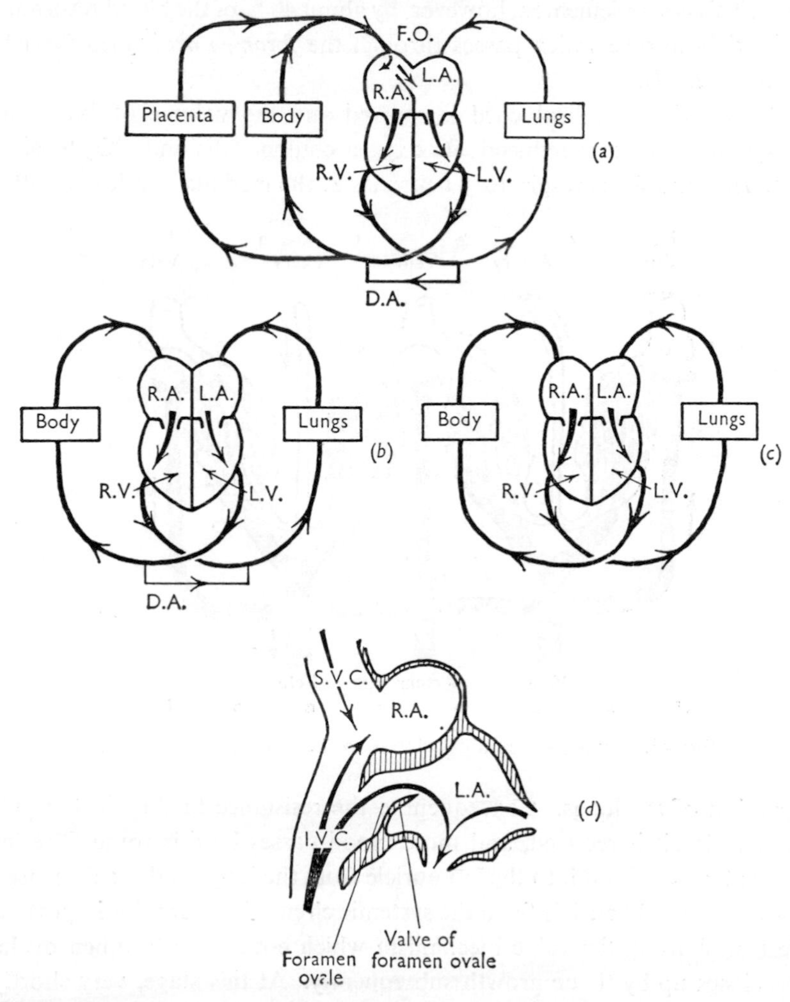

Fig. 47. Changes in the circulation during the development of the mammal. (*a*) Foetal, (*b*) neonatal, (*c*) adult, (*d*) foetal circulation, showing entry of vena cava to the right and left auricles.

to the means of gaseous exchange. The following account is based on the foetal lamb, but a similar change occurs in man. Of especial importance is the ductus arteriosus, which functions as a shunt allowing blood to pass from the pulmonary artery into the systemic circulation. In this way a greater volume of blood is passed to the placenta for oxygenation before its return to the right auricle. As so little blood circulates through the poorly developed lung capillaries, the left auricle receives only a small quantity of blood. This is supplemented, however, by about 40 % of the blood returning to the right auricle which passes through the *foramen ovale* into the left auricle (Fig. 47).

At birth the umbilical cord is severed and the volume of blood in the systemic circuit is reduced; its oxygen content falls and CO_2 tension rises. This stimulates respiratory neurones in the medulla and leads to the

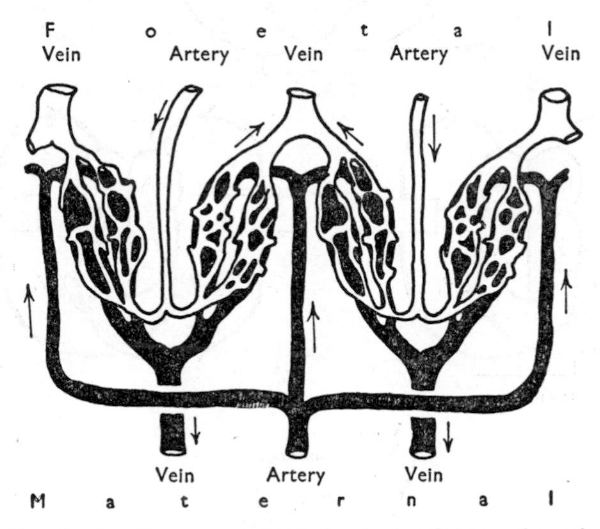

Fig. 48. Counter-current circulation in the placenta of a rodent.

expansion of the lungs. Consequently, the resistance to flow in the pulmonary circuit is reduced, and more blood passes by this route. The increased flow of blood into the left auricle from the lungs and the decreased flow into the right auricle from the systemic circulation have the important effect of shutting the valve mechanism which covers the foramen ovale. This closes up by tissue growth subsequently. At this stage, very shortly after birth, the ductus arteriosus remains open and in fact the direction of flow of blood through it reverses (Fig. 47 *b*). In this way a larger volume of the cardiac output passes to the pulmonary circuit than the systemic circuit and the relative inefficiency of the gaseous exchange in the lung is compensated for. Within the next 24 hr, the ductus arteriosus becomes occluded and the complete double circulation is established. The vital importance of the ductus and the foramen ovale are well known because of the effects resulting from their failure to close in development. Under these conditions the oxygenated and deoxygenated bloods do not remain separate and a blue baby may result.

In experiments with new-born lambs the importance of the ductus arteriosus has been demonstrated by occluding it very soon after birth with the result that the oxygen in the blood falls from 20 to 10 % saturation. Both the foramen ovale and ductus arteriosus act as shunts and play vital roles which tide the young mammal over the period during which the two circuits are incompletely developed and cannot take equal volumes of blood. Similar shunts seem to have persisted in different parts of the circulatory system of living amphibians and reptiles. We have already drawn attention to the ductus arteriosus in some amphibians, and a diagnostic character of reptiles is that the conus is divided into three from its origin with the ventricles. One of these is the pulmonary arch which divides into the right and left pulmonary arteries and the others are the right and left systemic arches (Fig. 54). In all living reptiles the left systemic arch takes some deoxygenated blood which becomes mixed with oxygenated blood when it joins the right systemic to form the dorsal aorta. This is clearly another mechanism whereby excess blood can be shunted from the pulmonary to the systemic circuit and other shunts are found within reptile hearts.

5.13 *The heart and circulatory systems of fishes, with particular reference to the dogfish*

The chordate circulatory system is characterised by the direction of flow being anterior in the ventral vessels and posterior in the main dorsal vessels. Blood is pumped forward from the ventral heart behind the gills into the ventral aorta from which afferent branchial vessels supply the gills. Oxygenated blood collects in the efferent vessels which communicate with the paired lateral dorsal aortae which convey blood backwards to the single dorsal aorta from whence it is distributed to the rest of the body. After its passage through the tissue capillary system it returns through small veins which lead into the main veins entering the auricle. This *single circulation* contrasts with that of mammals and birds in that all the blood must traverse at least two capillary systems before it returns to the heart. Consequently the pressure of blood supplying the tissues is lower than in the systemic circulation of birds and mammals. The circulation time has not been measured for many fishes but in an eel it may take 2 min and probably longer in dogfishes. Fish hearts, except in lung fishes, only

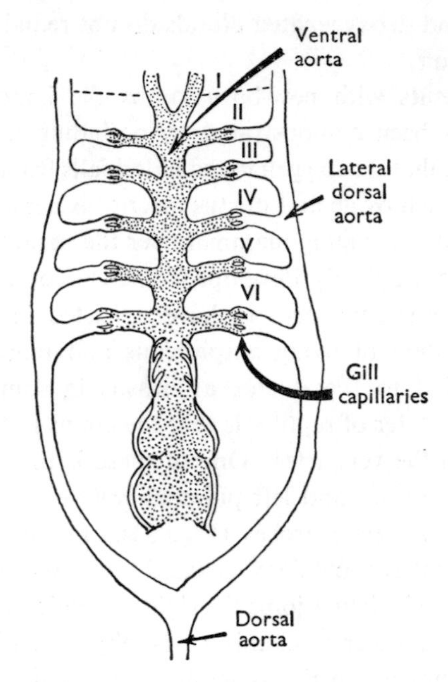

Fig. 49. Diagram of heart and arterial arches of a fish.

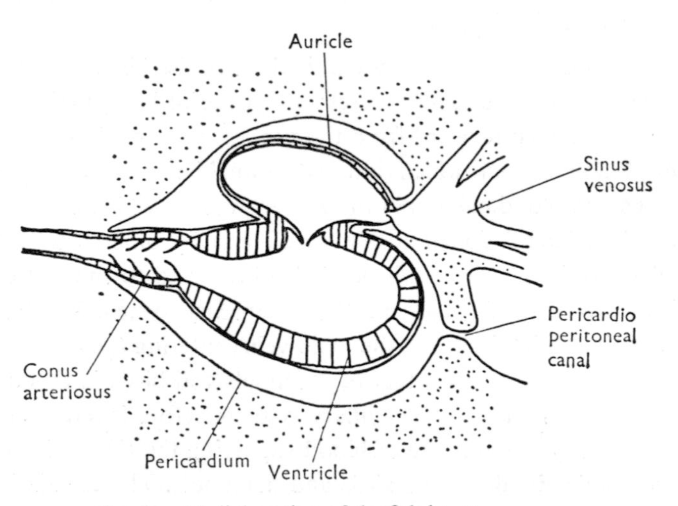

Fig. 50. Medial section of dogfish heart.

contain deoxygenated blood and receive their O_2 from a separate coronary circulation which takes its origin from the efferent side of the branchial circulation. The heart of a fish (e.g. dogfish, Fig. 50) functions as a single pump and is made up of four distinct chambers in series, each separated from the next by valves. The original ventral tube has become S-shaped as seen in side view. Blood from the venous sinuses collects into a sinus venosus from which it enters the auricle after passing the paired sinu-auricular valves. The auricle is dorsal and thin-walled and when it contracts the blood passes into the thick-walled ventricle. From the ventricle, blood is forced into the ventral aorta after having passed through a *conus arteriosus* or, in teleosts, a *bulbus arteriosus*. The conus is contractile, being made of cardiac muscle, and it has three longitudinal rows of semi-lunar valves, up to six in each row. In more advanced fishes, the conus becomes reduced but in all cases at least one pair of valves persists. The swelling in front of the ventricle in bony fishes is a fibrous non-contractile expansion at the base of the ventral aorta which lies outside the pericardium. The semi-lunar valves of the conus prevent reflux of blood into the ventricle during diastole. The bulbus, being elastic, evens out the pressure wave so that blood flows through the gill capillaries at a more uniform velocity.

5.131 *The dogfish circulation*. The heart is contained within the pericardial cavity, coelomic in origin, which, in the dogfish and other cartilaginous fishes, is contained within a box formed by the pectoral girdle. This is important functionally because, with the exception of a small duct (*pericardio-peritoneal*) which contains a valve, it is completely isolated from the main *perivisceral* (around the guts) coelom. Because of the single circulation and especially the large cross-sectional area of the venous sinuses, the problem of returning venous blood to the heart is very much more acute in a dogfish than mammals. Fishes do not possess valves in their veins although muscular activities must aid the venous return. The major factor assisting the return is the suction produced by the heart. This results from the incompressibility of the fluid contents of the pericardium which must maintain constant volume. When the ventricle contracts there will be a tendency for the volume to be reduced which can only be compensated by the flow of an equal volume of blood into the pericardium through the sinus venosus. Fluid cannot enter through the pericardio-peritoneal canal because its valve only allows the passage of fluid in the opposite direction. In this way the dogfish heart functions as an *aspiratory pump* which draws blood into itself from the venous sinuses.

Pressures within different parts of the circulatory system of the dogfish have been measured and shown to be much lower than those of mammals and other tetrapods. The systolic pressure within the ventral aorta is 23 mm Hg and falls to 15 mm Hg during diastole. This pulse pressure of 8 mm Hg is reduced after the blood's passage through the branchial capillaries (Fig. 51) and even further as it passes into the finer arteries. The pressure within the venous system is extremely low and in the large sinuses may even be negative. This is especially true in the region of the heart,

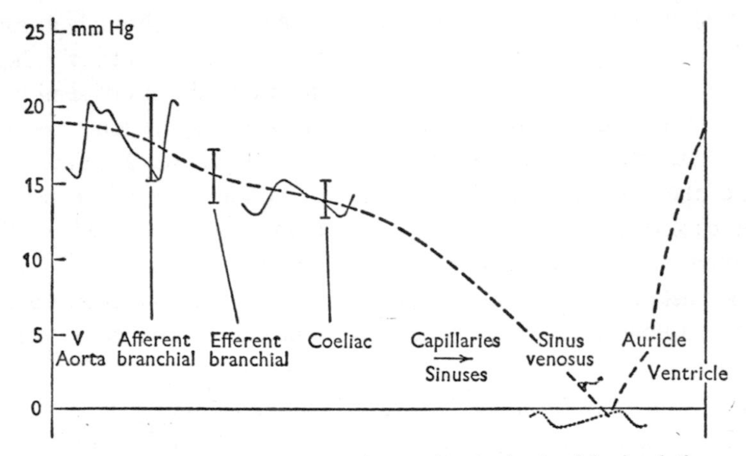

Fig. 51. Vascular pressures at various points in the dogfish circulation.

where blood in the sinuses will be subjected to the reduced pressure resulting from the aspiratory action of the heart. Circulation in bony fishes appears to be much more rapid and the pressures are higher at all parts of the system. The volume of blood in bony fishes is distinctly less than that of the dogfish, which is approximately the same as a mammal (70–80 ml/kg body weight).

Regulation of the frequency and output of the heart of fishes is controlled by the sympathetic and parasympathetic systems as in mammals. Changes in blood pressure are detected by receptors in the branchial arches in regions homologous with the carotid sinus of mammals.

5.14 *The heart of frogs, modern reptiles and birds*

The frog heart seems an ideal intermediate between the fish heart and that of mammal and bird for it contains a sinus venosus and two auricles but only a single undivided ventricle. Aortic arches 3, 4 and 6 persist as the carotid, systemic and pulmonary arteries. The evolution of the lung as an air-breathing organ was an important step in the conquest of the land. But it was equally important that mechanisms should evolve which enable the best use to be made of the oxygenated blood returning from the lung in the pulmonary veins. An essential stage which occurred during the evolution of all lung-breathing forms is that the auricle became divided, and the pulmonary veins returned blood to the left side. As in fishes blood from the rest of the body collects in the sinus venosus which opens into the right auricle. The subsequent fate of blood on the left and right sides varies in different tetrapods but some degree of separation is maintained in most of them. The main features of the classical (Brücke and Sabatier) description of separation in the frog heart are as follows:

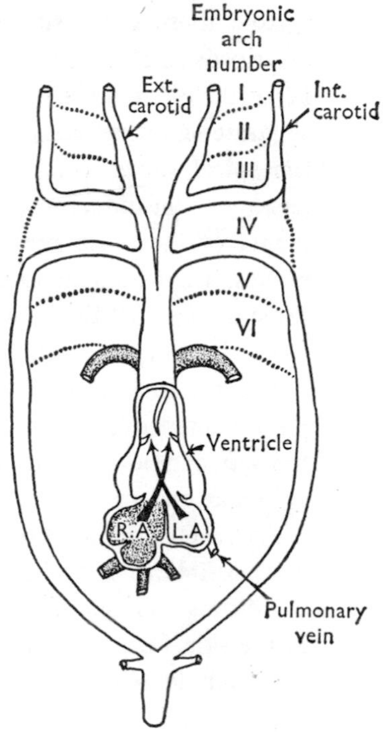

Fig. 52. Frog: heart and arterial arches.

(*a*) Separation is present in the ventricle because of its spongy nature and the high viscosity of the blood. Consequently at the completion of auricular systole, blood from the right auricle lies on the right side of the ventricle and that from the left auricle on the left side.

(*b*) When the ventricle contracts, the blood from the right side is the first to leave because it is from this region that the conus takes its origin. Successively passing up the conus arteriosus would therefore be deoxygenated blood, mixed blood, and lastly oxygenated blood from the left auricle.

(*c*) Because the resistance to flow in the pulmonary circuit is least the first lot of blood would flow into this channel. As the pressure rises in the pulmonary circulation, the second portion of blood (i.e. mixed blood) would pass into the systemic arch and only the last lot of blood (oxygenated) would be forced into the high-pressure system of the carotid arch. The high resistance of the carotid arches was thought to be due to the presence of the carotid bodies.

Some parts of this description have been established experimentally but others are now known to be incorrect, at least for the common frog

(*Rana*): (*a*) one feature which appears to be true is that there is little mixing in the ventricle. This has been demonstrated by injection experiments, so long as only small quantities (0·001 ml. Evans Blue) of dye are injected into the blood returning to the right or left auricle. (*b*) During ventricular systole there is no successive movement of the blood into the conus arteriosus as blood from the right and left sides pass up it simultaneously. Nevertheless, bloods from different parts of the ventricle maintain their separateness when passing up the conus, partly because of the action of the spiral valve. (*c*) Further evidence in support of the simultaneous passage of the different streams of blood up the conus comes from measurements of the pulse pressures which are simultaneous within the three arches. There appears to be little difference in the resistance to flow in the three arches. (*d*) Distribution of the blood into the aortic arches depends upon the morphological relationships of their exits at the anterior end of the conus. In *Rana* and *Bufo*, blood from the right auricle mainly enters the pulmo-cutaneous arch but some also enters the left systemic. Blood from the left auricle is distributed to the carotids and mainly to the right systemic, and never goes to the pulmo-cutaneous arch (Fig. 53).

As the O_2 content of the blood in the arches has not been measured, it is not possible to say whether there are significant differences between the carotid and systemic arches. It must be remembered that blood oxygenated at the skin forms an important proportion of the blood returning to the right auricle. Consequently, the difference in O_2 content of the right and left auricular blood may not be as great as might be imagined. In forms in which cutaneous respiration plays an even more important part than in *Rana* and *Bufo*, a greater degree of mixing of the bloods seems to occur, for example, in *Xenopus* the left auricular blood enters all three aortic arches. In urodeles a spiral valve is absent and complete mixing of the blood takes place in the conus. In some urodeles mixing even occurs in the auricle, which is undivided.

In reptiles there are two auricles, but division of the ventricle is incomplete in all groups except the crocodiles (the crocodiles are close descendants of the stock from which birds arose). Another interesting feature of the crocodile heart is that shortly after they leave the ventricles the left and right systemic arches communicate with one another through a small *foramen of Panizza*. The circulation of a bird can be derived from that of a crocodile by the loss of the left systemic arch. Derivation of the mammalian type of circulation is not possible from any living reptile. This is because the carotids of all modern reptiles arise from the right systemic

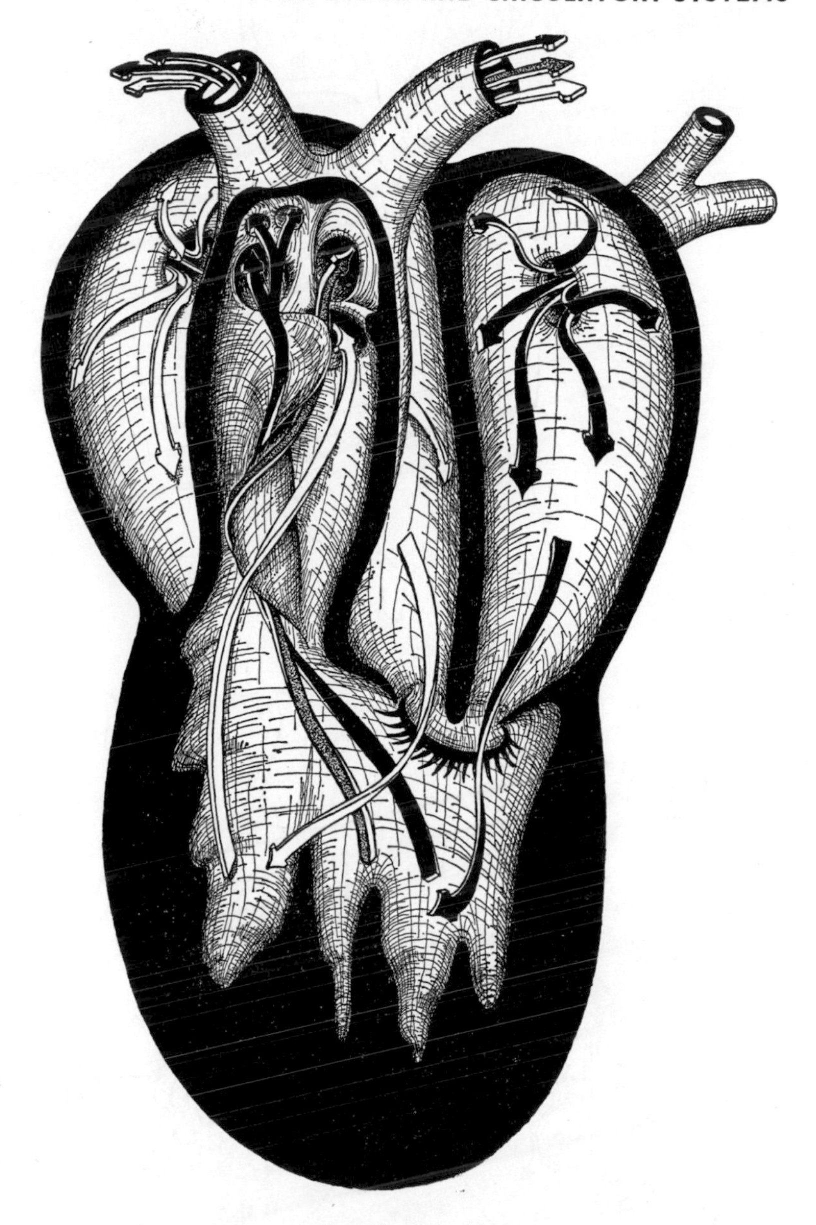

Fig. 53. Frog heart. Path of the blood indicated by injection experiments. Blood from the pulmonary vein (black arrow) passes to the carotid arches and the right systemic, blood from the vena cava (white arrow) enters the pulmo-cutaneous arch. Mixed blood (stippled arrow) passes to the left systemic.

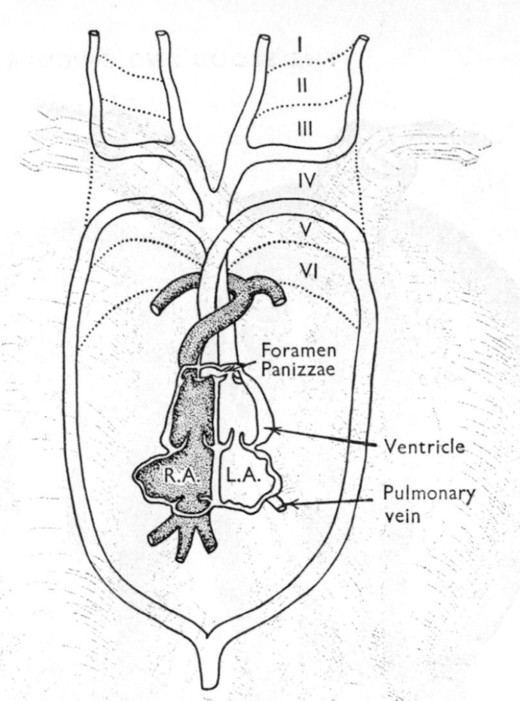

Fig. 54. Crocodile: heart and arterial arches.

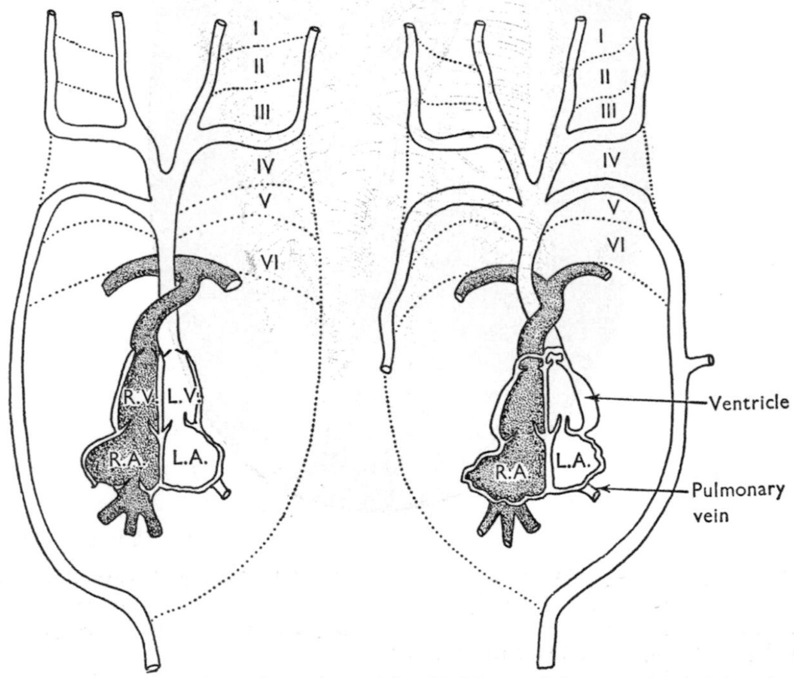

Fig. 55. Bird: heart and arterial arches. Fig. 56. Mammal: heart and arterial arches. (Note these are views with the heart pulled forwards so left and right appear to be transposed.)

arch, whereas in mammals they come from the persistent left systemic. Secondly, completion of the septum in any modern reptile except crocodiles would not result in a double circulation because both auricles open into the same part of the incompletely divided ventricle. Despite the possibilities for mixing apparent in these systems, injection experiments have shown that a precise distribution of the blood occurs, for example, in the lizard heart. Blood passing into the ventricle from the right auricle is distributed to the pulmonary arteries, whereas that which enters from the left auricle is pumped to the carotid and systemic arch. Mixed blood passes into the left systemic arch and becomes mixed with the right systemic blood when they both join to form the dorsal aorta.

5.15 *Phylogenetic considerations*

The persistence of so many apparently inefficient types of circulation in which the oxygenated and deoxygenated bloods are not kept completely separate suggests that they are related to some aspect of the functioning of the cardiovascular and respiratory systems. One suggestion is that the lungs have not evolved sufficiently to take a volume of blood equal to that passing through the systemic circulation. It is equally possible, however, that a greater volume of blood is passed through the pulmonary circulation in order to compensate for the relative inefficiency of the oxygen exchange mechanism.

The relationships between the different classes of vertebrates which can be deduced from studies of the heart and arterial arches are amply confirmed by reference to palaeontological data. The birds and crocodiles both evolved from the same group of reptiles—the Archosaurs. Other modern reptiles are on a different line of evolution but share with the crocodiles and bird ancestors the possession of two temporal vacuities (diapsid) in the skull. The reptiles which gave rise to the mammals were on an entirely separate line of evolution for they possessed only a single vacuity in the temporal region of their skulls. There are no living representatives of this group of reptiles (synapsids) and it is not surprising therefore that the mammalian type of circulation cannot be derived from any modern reptile. The circulation of the modern amphibians is doubtless specialised relative to that of the ancestral forms which first came on to land, but nevertheless it is from a type similar to this that the mammal circulation is most easily derived. The first stage in this evolution would probably have been the separation of the pulmonary arch from the systemic and carotid arches in its origin from the ventricle. At a later stage the right systemic arch would disappear and this would only have been possible had both carotid arches taken their origin from the left systemic arch. By whatever means it was achieved, however, the double circulation of both birds and mammals is a very efficient mechanism and ensures that blood from the left side of the heart which has been oxygenated at the lung is pumped to the body. Whether it goes to the left or right in the systemic arch is relatively immaterial.

5.2 The blood

The blood, lymph and coelomic fluids form the internal medium of the body, which is regulated in a precise way with respect to its ionic content, osmotic pressure, gas content, and temperature. This relative fixity of the internal medium is one of the most striking features of vertebrates, especially mammals and birds and, to a lesser extent, it is found in all animals. This tendency to maintain a constant internal environment is an excellent example of *homeostasis*. The efficiency of the controlling mechanisms is greatly aided by the rapid communication between different parts of the organism which the circulatory system provides.

5.21 *The structure of blood*

Blood is a tissue, in the same way that nerve and muscle are tissues; that is to say it is a collection of similar cells specialised to perform a given function in the body. Unlike most tissues, however, blood is a liquid and consists of solid corpuscles suspended in a fluid plasma.

5.211 *The red blood corpuscles* (erythrocytes). There are between five and six million erythrocytes in each cubic millimetre of blood and they are the most numerous type. The numbers fluctuate, but within small limits; certain organs, such as the spleen, act as stores and liberate more red corpuscles when they are required. In man the red corpuscles are biconcave discs which are $7{\cdot}2\mu$ in diameter with a thickness of $2{\cdot}2\mu$ and they have no nucleus. (This is a diagnostic character of mammals.)

Red corpuscles are formed in the marrow at the end of bones, in particular the ribs and vertebrae. They are proliferated, and of course destroyed, at a high and continuous rate (some 10^6/sec) and live for approximately 100–120 days in the circulation.

All types of corpuscle appear to develop from a particular sort of mother cell called a *haemocytoblast* and this goes through a number of stages including the reduction of the nucleus, which degenerates before the cell enters the circulation, and the addition of haemoglobin. For normal formation to take place, the constituents of haemoglobin must be present and the body requires inorganic iron as well as vitamin B and compounds containing cobalt and copper.

After their time in the circulation is completed the red cells are *phagocytosed* (engulfed) into the lining cells of spleen, liver, connective tissues and bones. The products of their breakdown are returned to the liver and some parts discharged into the bile caniculi from the sinuses of the latter. The breakdown product of haemoglobin is bilirubin ($C_{33}H_{36}N_4O_6$) and some of this is oxidised to biliverdin. Meanwhile most of the iron is released and stored in the liver for future use.

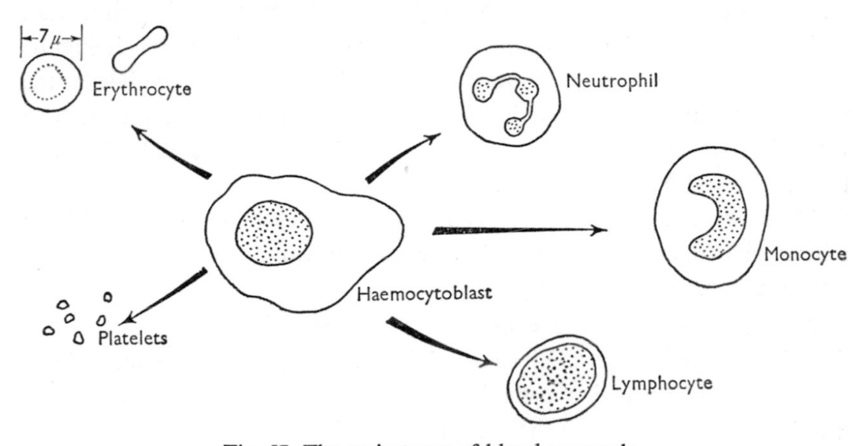

Fig. 57. The main types of blood corpuscle.

The functions of the red corpuscles are to transport oxygen, in the form of *oxyhaemoglobin*, and CO_2 as a *carbamino compound* with haemoglobin. The oxygenation of the haemoglobin also plays an important part in the release of CO_2 from the bicarbonate of the plasma (see p. 61). Finally it should be noticed that, despite their lack of a nucleus, red corpuscles are able to pump out any excess Na^+ and concentrate K^+ across their membranes.

5.212 *The white corpuscles* (leucocytes). These are found at only 5–6 thousand/mm^3, but like the red corpuscles their numbers, and in this case their types, vary considerably under different conditions. Leucocytes are produced variously by *reticulo-endothelial* cells in the liver, spleen, and lymph ducts, or by the bone marrow, and live for some two or three weeks in the circulation. Other than their white colour they can be distinguished from red corpuscles by the presence of a nucleus. As at least five types of white corpuscle are found it would be as well to deal with the appearance and properties of each type separately. In general they all play a part in the defence of the body against infection.

Neutrophils (or polymorphs) are distinguished by their lobed nuclei; they are phagocytes, and make up 70 % of all the white corpuscles. These neutrophils will be attracted out of the general circulation at any site of infection and despite the fact that they are larger than red corpuscles they are able to pass through the walls of the capillaries by passing through a thin canal of protoplasm into a *pseudopodium* (protoplasmic process) on the far side. (This is called *diapedesis.*) They secrete certain enzymes but there is no doubt that their effectiveness as phagocytes is helped by secretions of antibodies from other types of white cells (the lymphocytes).

Once the neutrophil has found bacteria or other foreign bodies it engulfs them in the same way that amoeba takes in its food and, once inside, the invading organism is digested. Many neutrophils, however, succumb to the poisons (toxins) of the germs and are killed, forming the bulk of the pus that collects about an infection.

Monocytes are larger than red cells, being 10–15 μ in diameter, and have a bean-shaped nucleus. They act in conjunction with neutrophils, described above, and are strongly phagocytic.

Lymphocytes are white cells that have entered the blood from the *lymphatic system* from which they originate. Their function seems to be to provide *globulin* proteins in the blood from which *antibodies* can be made and like the first two types they collect at the site of bacterial invasions. The nucleus is spherical and they are not phagocytic (more will be said about these under lymph (p. 107)).

The remaining types of white corpuscle are *basophils* and *eosinophils* and these are normally few and far between in the blood. Not much is known about them except that the latter increase in certain types of infection and allergic conditions. They both have lobed nuclei.

The integration of the activity of these types of corpuscle and their place in the defence of the body are best considered when the reticulo-endothelial system and the lymph have been described.

5.213 *The platelets.* These are the final solid constituents of the blood and are about one-quarter of the size of the red corpuscles. Their approximate density is 250,000 per c.c. and they are formed by the disintegration of very large cells, the *megakaryocytes* of the *red bone marrow.* Like the red corpuscles the platelets have no nucleus and consist only of lumps of granular cytoplasm.

Platelets are concerned with the *clotting* of blood, which is important in wound healing and part of the body defence mechanisms. This depends on the production of an insoluble protein, *fibrin*, in the blood in which the corpuscles get entangled.

The mechanism of clotting is as follows:

(*a*) On exposure to air platelets produce a substance *thrombokinase* (or thromboplastin). This can also be made in wound tissues.

(*b*) This substance, in the presence of calcium salts, acts on the precursor of an enzyme, *prothrombin*, and activates this to *thrombin*. Prothrombin is found as a soluble constituent of blood plasma.

(*c*) The thrombin now causes soluble *fibrinogen*, also present in the blood, to turn to insoluble fibrin and this is the basis of a clot meshwork. The time to form a sealing clot varies from 6½ to 20 min. Thus:

$$\left.\begin{array}{l}\text{Foreign bodies,}\\ \text{air or wounding}\end{array}\right\} \rightarrow \text{Platelets} \rightarrow \text{thrombokinase}$$

$$\text{Thrombokinase} + \text{Ca}^{++}\text{ salts} + \text{prothrombin} \rightarrow \text{thrombin}$$

$$\text{Thrombin} + \text{fibrinogen} \rightarrow \textit{fibrin clot}$$

The strands of fibrin can be seen radiating out from the platelets as the blood clots and the whole mechanism takes place very rapidly. It will not take place inside the circulation as there is no stimulus to activate prothrombin, and if, by chance, small quantities of thrombin are formed they are destroyed by a special antithrombin in the plasma. Prothrombin needs a supply of vitamin K in the diet for its formation.

5.214 *The plasma.* A straw-coloured fluid making up about 50 % of the blood volume. It contains 90 % water and 10 % dissolved substances, including proteins, food substances, salts and excretory products.

Proteins in the plasma make up most of this solid material and have a number of specific functions. The main protein is *albumen*, while others include globulin and fibrinogen, and all of these combine to produce a colloid osmotic pressure (about 30–40 mm Hg) which in turn helps to control blood volume and water balance in the body. The globulin is concerned with the synthesis of antibodies in the blood and is probably released by lymphocytes and reticulo-endothelial cells, while the function of the fibrinogen has been described above (i.e. formation of the fibrin clot). These proteins are all manufactured in the body and may be differentiated from the small percentage of amino acids being transported by the blood; for example, from the ileum to the liver and from the latter to other tissues.

Besides the amino acids, fatty acids and sugar (in the form of monosaccharides) are transported in the plasma from the gut to the liver and thence to all parts of the body. There are some 0·6 g fat and 0·1 g sugars in 100 c.c. blood, and the percentage is strictly regulated because outside a certain range of concentrations these substances have deleterious effects on the body, the former on pH and the latter osmotic. The main mechanism controlling the blood sugar depends on insulin secreted by the pancreas (p. 233).

Other organic substances are excretory *urea*, *creatinine* and *uric acid*, which are found collectively at concentrations of 0·3 g/100 c.c. blood, together with traces of lactate and hormones.

The inorganic ions include K^+ 5 mM/kg, Na^+ 143, Cl' 103, Mg^{++} 2·2, Ca^{++} 5, together with traces of sulphate, phosphate and carbonate.

The functions of these ions have been described elsewhere (p. 125), but the alkali present has an important buffering function and together with the haemoglobin and plasma proteins forms the alkali reserve of the body. From the physiological point of view the most important of these buffers is the bicarbonate because the nervous mechanisms regulating respiration are tuned to maintain a constant blood pH or CO_2 content.

As a chemical buffer bicarbonate mops up hydrogen ions from organic acids as follows:

$$CH_3CHOH.COOH \rightarrow CH_3CHOH.COO' + H^+$$

(lactic acid)

(Sodium and bicarbonate ions in blood) → Na^+ HCO_3'

$$CH_3CHOH.COONa \quad H_2CO_3 \rightarrow H_2O + CO_2$$

But phosphates and haemoglobin in the red corpuscles and plasma proteins are also important chemical buffering systems.

A small amount of oxygen (0·25 %) and carbon dioxide (3 %) are carried in solution in the plasma, but without haemoglobin this would not be adequate for the needs of the mammal.

5.215 *Exchange between the capillaries and the tissues.* The blood, with the composition described above, passes into the *capillaries* from the small arterioles of the arterial system. These tiny vessels (some 8–10 μ in diameter) permeate all the tissues of the body. It is across the capillary walls that the cells of the organism must take up the substances they require.

In order to understand how these exchanges take place it is necessary to visualise the arterial blood entering the capillaries at a hydrostatic pressure of some 44 cm H_2O, which has the effect of pushing out small-sized molecules through the permeable walls of the capillaries. Across these walls O_2 and CO_2 can also pass, and white corpuscles may leave the capillaries as required. At the distal end of the capillary network, the hydrostatic pressure is much reduced (only 17 cm H_2O) and the osmotic pressure of the blood proteins, which do not pass across the capillary walls, tends to draw back some of the water molecules that have escaped from the capillary. There is thus a circulation of the blood plasma as the tissue fluid in the capillary networks but not all of it is returned to the capillaries at their venous end. This excess filtrate forms the lymph.

5.216 *The lymph.* As described above, many substances leave the capillaries and enter the tissue spaces. The fluid thus formed is called lymph and is in immediate contact with the cells of the tissues. Lymph has much the same composition as the blood plasma but lacks red corpuscles and has various other minor differences. These are as shown in Table 5:

Table 5

Substance	Lymph (as compared with plasma)
Ca^{++}	Less
PO_4'''	Less
K^+	Less (by 3%)
Protein	Less (by 38%)
Fat	Less
Glucose	More (about 20%)

The lymph has more lymphocytes, which originate in the lymphatic system, but less of the other type of white corpuscles, and it passes from the tissue spaces into the small vessels called lymphatics. These lead into larger vessels which eventually drain into a tube called the *cisterna chyli* (Fig. 58). This tube passes into the venous system just as the latter enters the right auricle, that is, at the base of the jugular vein, by the *thoracic duct.* The blood pressure here is only 5 mm Hg, which is less than the lymph pressure, and thus the lymph fluid returns to the general circulation. Lymph vessels are similar to the veins and depend for their flow on muscular activity and valves which prevent backward movement of the fluid. At various points along these vessels are *lymph nodes,* essentially filters, concerned with the defence mechanisms of the body.

These nodes are prominent in the intestines and in certain regions such as the back, groin, and under the arm. The lymph returning to the heart enters the node in a sort of pool—it then passes into the body of the node where great masses of lymphocytes are proliferated from the reticulo-

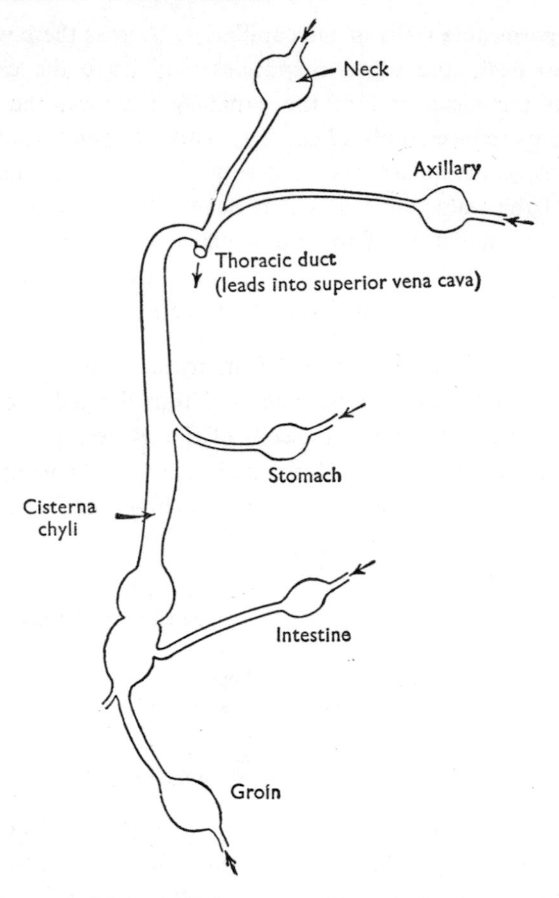

Fig. 58. Layout of the lymphatic system. (The round areas represent regions where many nodes are found.)

endothelial system (see below), and thus into the efferent vessel. It is known from experiments on dogs that a lymph node is able to remove great numbers of foreign bodies such as germs from the circulating fluid and this is their main function. In cases of infection the lymph nodes (or glands) become swollen.

These nodes are only found in warm-blooded vertebrates.

5.22 *The reticulo-endothelial system*

The cells of this system line the *sinuses* of the liver, spleen, lymph and connective tissues of the body. They are partly sessile or fixed and partly mobile and are concerned in the defences of the body. Sessile members are able to ingest foreign matter as it passes and to destroy it while the large macrophages form the mobile cells, which engulf germs and ingest them. The cells of the reticulo-endothelial system also produce antibodies and *anti-*

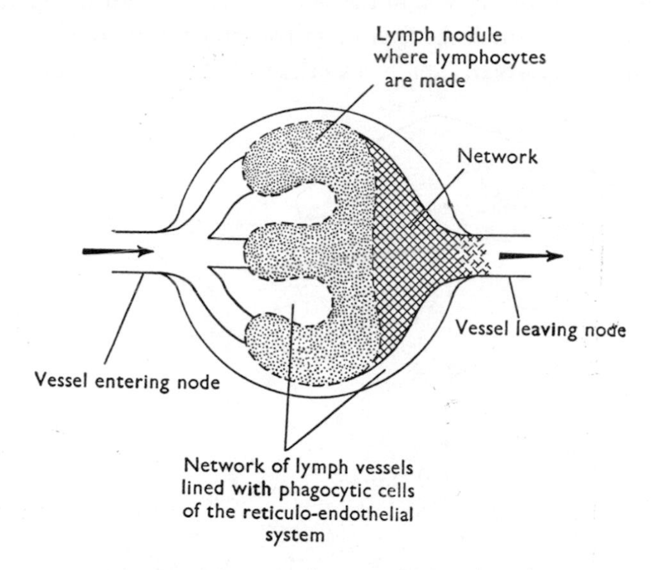

Fig. 59. Schematic diagram of a lymph node.

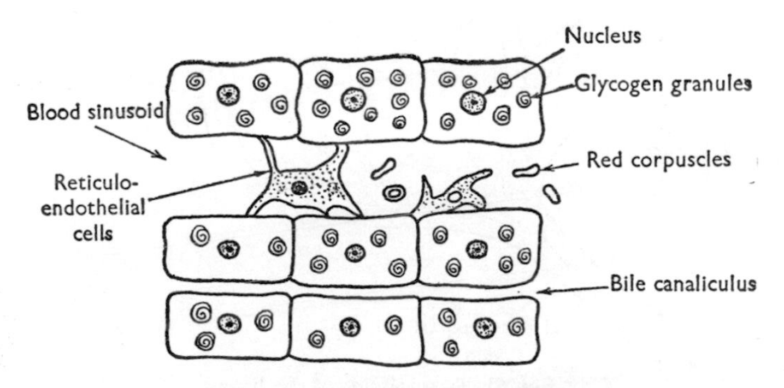

Fig. 60. Enlarged part of blood sinus of liver.

toxins in response to specific stimuli and these circulate in the body fluids acting against foreign bodies. The monocytes already described are budded off from the reticulo-endothelial system and one of the normal functions of this system is the destruction of old red blood cells.

5.23 *The liver*

The liver is an important organ of the body vitally concerned with the regulation of substances within the bloodstream. As all the blood of the mammal passes through the liver every two minutes or so, changes taking place in this large organ rapidly affect the whole circulation.

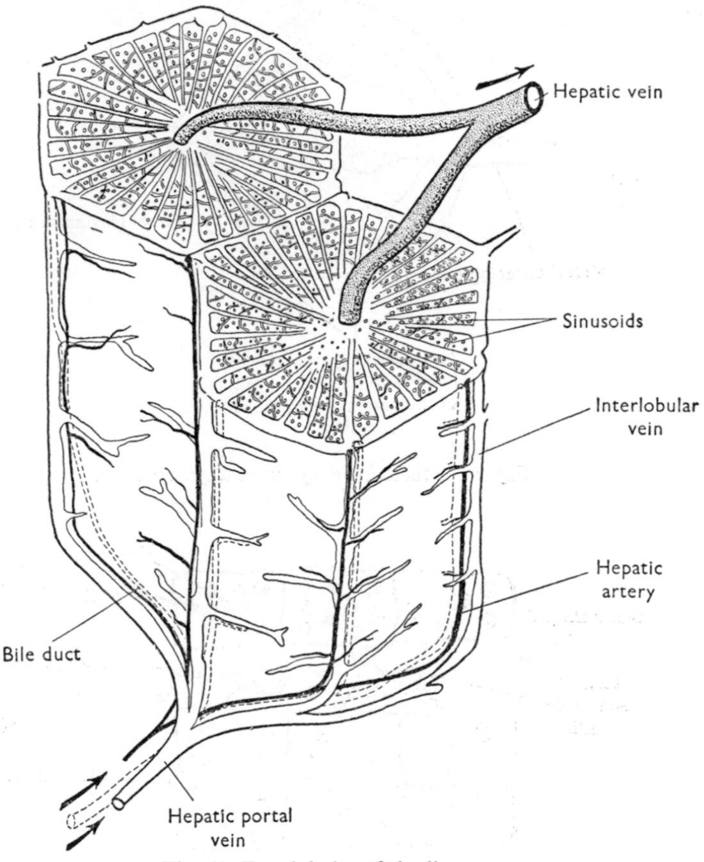

Fig. 61. Two lobules of the liver.

The liver takes up one-fifth of the whole viscera and is composed of four lobes and has a uniform histology, being made up of large numbers of similar *lobules*. These tend to be hexagonal and from a central vessel radiate out 50 or so small sinusoids along which blood passes from the peripheral vessel. It is in its passage through these sinuses that exchanges take place (Fig. 61).

As with other organs, the liver receives an arterial supply (the hepatic artery from the aorta) and drains into a vein, the *hepatic*, which passes into the *posterior* (or *inferior*) *vena cava.* In addition to this the liver also receives the terminal branches of the portal system from the ileum and other parts of the intestines. Detailed arrangement of these three systems in relation to the lobule is shown in Fig. 61. Branches of the hepatic portal system pass into the edges of the lobule while the hepatic vein originates from its centre. Hepatic arteries are also at the periphery.

Besides the vascular system the liver has a complex meshwork of bile canals leading from the intercellular canaliculi and entering the *gall bladder*.

The functions of the liver are as follows:

(*a*) The reception and storage of food substances assimilated in the intestines.

(*b*) The deamination of proteins.

(*c*) The production of bile.

(*d*) The removal of waste substances and foreign matters and general detoxication of the blood.

(*e*) The production of heat.

It therefore completes and integrates many of the metabolic processes that have already been described, such as digestion.

The reception and storage of food is by uptake from the hepatic portal vein of substances absorbed from the ileum. Monosaccharides, usually in the form of glucose, are absorbed along cells of the liver sinusoid and condensed into glycogen. This is a polysaccharide similar to starch and the cells of a normal liver are full of glycogen granules. This condensation involves a phosphorylation and requires energy. It is stimulated, both in the liver and the tissues, by the action of insulin, which seems to facilitate the passage of the sugar across the cell walls. Glycogen is hydrolysed back to glucose and passed into the blood, where a constant level is maintained under the influence of hormones secreted by the anterior lobe of the pituitary, pancreatic islets and the adrenal cortex and medulla. It is thus transported all over the body and can be taken up by cells according to their requirements.

Fats and fatty acids entering the liver, either by the hepatic portal system or the general circulation, are metabolised in the liver. Much fat is stored in its cells and the breakdown and respiration of fat (p. 273) also takes place in the liver. Essential amino acids may pass through the liver into the circulation and thence be taken up by tissues. Non-essential and excess amino acids are taken up by the liver cells for deamination and subsequent respiration. This deamination is an important process and consists of the removal of nitrogen from excess amino acids. The nitrogen in the form of ammonia is converted to urea by reacting with carbon dioxide (see ornithine cycle, p. 273):

$$2NH_3 + CO_2 \rightarrow CO(NH_2)_2 + H_2O.$$

Besides the removal of nitrogen in the liver, other constituents of amino acids, including phosphorous and sulphur, are also removed and eliminated as waste materials, leaving carbon, hydrogen and oxygen compounds.

The mechanism of bile production has been described on p. 33. The bile canaliculi run parallel with the sinusoid of the liver and bile substances are continually passed into the canaliculi from the surrounding cells. At the edges of the lobule are found the longer bile ducts which eventually empty into the gall bladder.

Part of the reticulo-endothelial system (see above) lines the sinuses of the liver lobules and this gives the liver the ability to pick up and destroy foreign bodies. More than this, these cells in the liver seem to have a wide range of breakdown effects and are able to down-grade and excrete a number of toxic substances taken into the bloodstream, for example, alcohol.

Finally, it is clear that the liver is an organ of very high metabolic activity in which many syntheses are made (e.g. blood proteins), and under certain conditions it is able to increase its heat production and this heat is transferred to the rest of the body in the general circulation.

5.24 *The defences of the body against infection*

5.241 *The causative agents.* Disease is a state of deleterious departure from the normal condition of the body. This is a wide definition and covers *genetical abnormalities* of form and function, sometimes called 'inborn errors of metabolism', *functional diseases*, and the presence of parasitic organisms. It also covers the *deficiency diseases* which result from the body being deprived of some essential constituent of diet. It is only the defences of the body against infectious agents that concern us here and the role of the blood and reticulo-endothelial system.

The infectious agents of disease fall into the classes of *viruses*, *bacteria*, *protozoa* and the variety of *macro-parasites*. Examples of these as well as of other diseased states are shown below (Table 6).

Viruses are entirely parasitic and can only lead a normal life within the body of their host, but as far as most micro-organisms are concerned the body is by no means an ideal environment and it is not surprising that so few of the many thousands of species of bacteria or protozoa have been able to live within the tissues. On the whole, the mammalian body is too hot for the most efficient functioning of micro-organisms (hence the old-fashioned method of curing syphilis by giving the patient malaria which raised the body temperature and killed the syphilis bacteria) and the chemical environment of the body is inimical to their growth. It is not practical for a parasite to destroy its host, and the usual state that exists between parasite and host is one of adaptation rather than violent reaction. We all carry great numbers of potentially pathogenic bacteria on our bodies and on the linings of the respiratory tracts as well as in the gut. Should our defence mechanisms fail these organisms would multiply and bring about disease, as happened with the enteric infections at Hiroshima when radiation temporarily suspended the natural relining of the gut by cell divisions.

Table 6. *Human diseases due to micro-organisms, larger parasites and other causes*

Bacteria	
Pneumonia	Dysentery (one type)
Tetanus	Boils
Typhoid	Tuberculosis
Diphtheria	Scarlet fever
Cholera	
Virus	
Mumps	Rabies
Influenza	Yellow fever
Smallpox	Poliomyelitis
Measles	
Protozoa	
Dysentery (amoebic)	Sleeping sickness
Malaria	
Other parasites	
Elephantiasis (nematode)	Liver fluke (platyhelminth)
Bilharzia (platyhelminth)	Tinea (fungus)
Diseases due to deficiency of necessary food substances	
Simple goitre	Beri-beri
Scurvy	Pellagra
Rickets	
Functional diseases due to failure of organs	
Heart conditions—angina, etc.	

In order to multiply, the infectious micro-organism must first enter the tissues of the body, and this is not easy as the skin limits such entry by the continuous sloughing off of dead keratinised epidermis. Inside the body the various tracts are protected by wandering white cells and by secretions, both of which destroy infectious agents as well as having a commensal flora which resists the settling of new species. The nose filters out foreign matter from the air taken into the body and the acidity of the stomach does much to limit bacteria entering the small intestine. In the reproductive and respiratory tracts cilia beat towards the openings of the body and pass up particles, which are expelled. If the body is wounded a blood clot seals off the tissues from micro-organisms.

In other words, it is neither easy for a pathogen to enter nor to survive within the body and this is why we remain relatively free of disease most of the time.

5.242 *The disease-causing micro-organisms.* (*a*) *Viruses.* The virus is a microscopic entity some 12–400 mμ, made of *nucleoprotein* and usually surrounded by a protein wall. It has no internal organelles and although it possesses a genetic mechanism, indeed consists of a genetic mechanism by which it can reproduce, it lacks an energy-exchange mechanism. For this reason it must be parasitic, using the energy-exchange units of its host cell to satisfy its own metabolic requirements.

Viruses attack the cells of their host by injection of the nucleoprotein part into the *cytoplasm*, the protein case remaining behind. The core of the virus rapidly reproduces, taking over the genetic control of the host and using it to synthesise its own nucleoprotein. In a matter of minutes the host cell may be completely converted to virus material, giving an increase in the latter of some 200 times, and this new virus is released to attack further cells. Sometimes viruses remain latent in a cell, only becoming active when subjected to abnormal conditions such as heat or radiation. An example in man is the skin infection *herpes simplex*, which is usually carried as a latent virus. From the above it can be seen that the virus represents an extreme reduction of the life-form towards parasitism, and against these organisms the body has to depend on its own defence mechanisms as *antibiotics* are not effective against viruses. In the mammal, viruses cause infections of the skin, nervous tissue, respiratory tract, alimentary canal, as well as general disorders (see Table 7).

(*b*) *Bacteria.* These are unicellular organisms ranging from some 1 to 8 μ in size, although exceptional forms may be much larger than this, which contain a type of nucleus and organelles. They have a form of sexual reproduction and a wall of *chitinous polysaccharide* resembling the fungi. In adverse conditions the bacteria tend to form resistant spores which remain alive and potentially infective for long periods.

Bacteria fall into three classes. First there are the round *cocci* which may be subdivided into single cell types (such as the pneumonia bacillus), chain forms (*streptococci*) and cluster-forming species (*staphylococci*). The second type of bacteria is the rod-shaped *bacillus*, and the remaining type the spiral-shaped *spirochaetes*.

Table 7. *Viruses and man*

Tissue affected	Disease	Transmission
Skin and mucous membranes	Smallpox	Contact, sputum, etc.
	Measles	Contact
Nervous tissue	Poliomyelitis	Faeces, sputum
	Rabies	Animal bites
Respiratory tract	Influenza	Nasal and oral discharges
	Pneumonia	Sputum
Intestinal	Epidemic diarrhoea	Faeces
General infections	Yellow fever	Mosquitoes
	Dengue	Mosquitoes

When the pathogenic forms invade the body they may cause damage by the destruction of cells or by their accumulation, as in diphtheria. It is, however, far more usual for the harmful effects to be due to the metabolic wastes or toxins that they produce. Some of these toxins are among the most potent poisons known, thus some 0·0001 c.c. of botulin toxin can cause the death of a guinea-pig. In some cases, such as tetanus and gangrene, the bacteria feed on living cells and dead wound tissues but the powerful toxins they produce diffuse into uninfected tissues and cause death.

In nature, bacteria compete with many other micro-organisms and some of these produce substances which destroy the bacteria. Well-known examples are the fungal extracts penicillin and streptomycin, and these can be used to combat infectious bacteria in the body.

5.243 *The production of antibodies.* It is part of the functions of living systems to produce specific substances against foreign proteins and neutralise their effects. It has been repeatedly shown that the chemical system of each individual is unique, and this is only to be expected from the genetic mechanisms of reproduction. Thus, to the individual, foreign protein includes not just micro-organisms but tissues, such as blood, from another member of the species.

The basis of immunity against the types of diseases we have been considering is chemical—the release, or in the case of virus infections, formation within the cell, of a specific substance that will destroy the foreign body. The substances released by an organism for this purpose are called antibodies and where they are concerned with neutralisation of toxic excretions of infectious agents they are called antitoxins. Substances which bring about the liberation of antibodies are termed *antigens*.

(*a*) *The nature of antibodies.* Antibodies are produced by the cells of the body and the local inflammation which follows infection is due to the secretion of chemicals from the affected cells. However, the specialisation of organ systems has led, in the mammals, to the elaboration of particular tissues for antibody production. Such tissues are the wandering lymphocytes and the sessile reticulo-endothelial system of the blood and lymph, as well as the wandering phagocytic cells (sometimes called macrophages).

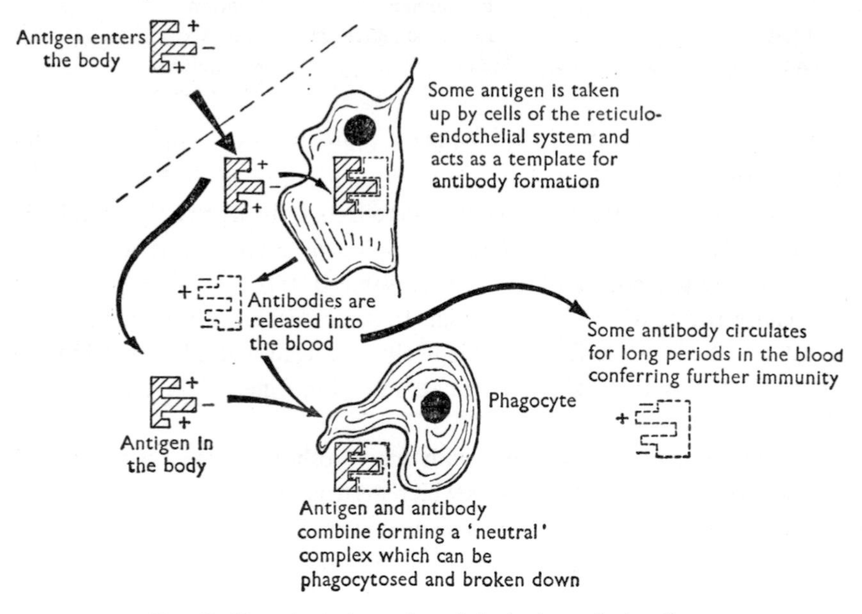

Fig. 62. Hypothetical reaction of the body against antigens.

Chemically the antibodies belong to the class of soluble proteins called globulins and the active antibody has some specific modification whereby it can react with its antigen. Production of artificial antibodies shows that the antigen causes the antibody molecule to produce specific chemical groups which react with many surface groups of the former. The properties of antibodies are those of proteins and they can be *denatured* and destroyed by protein-altering agents. Their molecular weights range from 100,000 to 1,000,000—the individual producing molecules in one size grouping. The antibody is made in quantity after contact with the antigen, a period of one week usually representing the period of maximum production. Once produced the antibody may circulate in the blood and other body fluids for very long periods—even years; synthesis of such an antibody is also more rapid on any subsequent infection. It is almost as if the cells of the body remembered the chemical pathway involved. An alternative suggestion is the 'clone theory' of antibody formation where certain endothelial cell clones are selected which are able to produce a specific antibody. These clones tend to survive, and their daughter cells retain the ability to secrete the original antibody.

While it is not exactly clear how antibodies are formed, the most acceptable theory is that the antigenic molecule itself directs their production so that the configuration of the former's side-chains fits the template provided by the antigen.

(*b*) *The reaction between antibody and antigen.* Antibodies may react with antigens in various ways. The antigen may become agglutinated (insoluble) or suffer lysis (chemical breakdown). Opsoninisation is where the antibody affects the surface of the antigen to allow it to be more readily engulfed by phagocytes.

It seems likely that the whole molecules of antibody and antigen react together at their surfaces so that respective groups couple exactly and a large neutral complex is made. The whole chemistry of the antibody, as described above, has been 'tailored' to fit on to and neutralise the active surface of the antigen. The neutral combination of antigen and antibody is destroyed by enzyme action or by phagocytes. Particularly important antibody groups are Cl' and CH_3 and these tend to react with COO' and NH_3 on the antigen.

In lysis, the dissolution of an antigenic cell, the antibody combines with the cell in small quantities (some 30 molecules of antibody are able to destroy a single red corpuscle in man), and in order to do this a further immunological substance called *complement* is required. This plasma substance complement, through its enzymic activity, dissolves the antibody–antigen combination on the surface of the cell, thus allowing the contents to escape, but for this lysis to happen both complement and antibody work together.

5.244 *The action of phagocytes.* The part played by the phagocytic white cells in the defence of the body is the engulfing of foreign material and this is done much more efficiently when the correct antibody is present. As already described, the phagocytes fall into those that wander through the blood, lymph and tissues, and those that are sessile reticulo-endothelial cells. The free phagocytes are rapidly attracted by the presence of antigens which are engulfed and digested. Should such foreign bodies enter the deeper tissues or the blood they will be passed through the lymph nodes, the liver and other structures where reticulo-endothelial cells are found and there they will be taken up and destroyed. Once the phagocyte has consumed a certain number of bacteria it dies and itself forms a constituent of the pus which is made at the site of an infection. Pus is also the result of dead body cells killed by toxins, of dead bacteria and other debris.

If an infection is kept local, connective tissue grows round the wound site, which is isolated above the new healthy cells, any pus formed being expelled by the elasticity of the connective tissue. A more generalised infection leads to phagocytic action throughout the body, although the results of such action may be transported to the surface, where they are eliminated.

5.245 *The role of antitoxins.* Antitoxins are special types of antibodies but act against the toxins formed by infectious agents, which they neutralise. They act in the same way as the antibodies and are made by the same sort of cells that produce the latter.

5.246 *Protection against virus infections.* The virus attacks inside the cell and therefore is not readily accessible to the circulating antibodies or the phagocytes, or indeed to chemicals, such as the antibiotics, introduced into the body.

In 1957 it was shown that a cell produces against the virus that attacks it a substance, now called *interferon*, which seems to prevent the virus making use of the energy mechanism of the cell. (It will be remembered that the virus needs to use the cell's energy supply system in order to reproduce.) This interferon has a very similar composition and molecular weight wherever it is found and there is some hope it may be possible to produce it for clinical use. Of course the mammal also makes antibodies against viruses and these appear before interferon and are an essential part of the body's defence.

5.247 *Natural immunity and active and passive immunisation.* The defence system of the body as outlined above is stimulated to produce antibodies not only when a disease is present but under a number of other conditions. Such a production of antibodies may give rise to natural immunity; that is, an immunity to a disease which has not been acquired by medical means. During life the body is exposed to many potentially harmful micro-organisms, such as the polio virus or tuberculosis bacteria, which may enter the body and stimulate its antibody system. The latter can resist and destroy the infectious agent, perhaps without any clinical symptoms of the disease appearing, and the antibodies formed provide further immunity against subsequent infections. In other cases the disease will manifest itself (e.g. mumps, measles, whooping cough) and on recovery the body retains antibodies, as well as the 'memory' of their synthesis, which will provide a lifelong *natural immunity*. Viruses exist in more mutant forms than bacteria and this is one of the reasons why prolonged immunity to such diseases as colds and influenza does not occur.

Artificial immunity consists in the formation of antibodies to a disease when this process has been stimulated by the deliberate introduction of antigenic substances into the body. In some cases weakened or related micro-organisms are introduced into the body (for example, rabies, smallpox immunisations), in others toxoids (weakened toxins) are introduced (e.g. tetanus). Dead antigenic material also stimulates antibody production without fear of infection. Such a form of immunisation is called *active* because it relies on the body's own defence mechanisms and in most cases leads to prolonged immunity. This whole system of artificial immunisation is usually attributed to Jenner, who 200 years ago showed that the giving of the mild disease cowpox prevented catching the much more serious disease smallpox.

Another form of artificial immunity is called *passive* and this is used where an infection has already taken place. Well-known examples are the tetanus antitoxin given after a deep wound and the snake-bite antitoxin administered after a bite. These antitoxins are prepared by the inoculation of an animal, such as a horse, with increasing doses of antigenic material and collecting the antitoxins it forms. The method is passive because, although it assists the body to neutralise the infection, it does not stimulate the cells to produce antibodies and thus does not confer more than a temporary resistance.

By the use of these techniques, together with public health measures and the use of antibiotics, the incidence of bacterial diseases in many parts of the world has decreased. That such diseases are only kept under control and not permanently destroyed is shown in places where our modern health measures have temporarily collapsed, as in the Congo in 1959, and in India in 1949 when, during the pilgrimage to the Ganges, an epidemic of smallpox broke out owing to the failure to ensure a widespread programme of immunisation.

5.248 *Antibody reactions not associated with micro-organisms.* There are a number of ways in which foreign protein may be introduced into the body other than by micro-organisms. Such foreign proteins can evoke antibody responses by the body which are not only of importance to the physician but also illustrate the very wide scope of protective reactions.

Many individuals are sensitive to a variety of animal and plant proteins which produce *allergic responses* in the cells with which they come in contact. Grass and other plant pollens may stimulate antibodies and increase histamine secretion (a hormone involved in defence mechanisms) in the cells of the respiratory tract, eye membranes, etc., a condition known as hay fever. Other types of protein may cause asthmatic attacks. A serious type of allergy is where the body has previously been made hypersensitive to a protein, for example, the venom of a bee, and a subsequent dose causes a general reaction in the tissues. This exaggerated response can lead to the death of the individual, but despite its disadvantageous nature it is only an extreme form of antibody action.

Another manifestation of the sensitivity of the body to foreign protein occurs in blood transfusions. Certain factors in the corpuscles can act as antigens when transfused into the body of an animal with a different genetic complex. For man these antigen–antibody pairs are most typically found in the A, B, AB, O groups. The large letters (except for the O) represent antigens and are clotted by the opposite antibody; that is, α, β, $\alpha\beta$. Thus A blood is clotted by both B and O blood, but not by A or AB groups, while O blood can be transfused safely into any blood group. This transfer of protein from one individual to another does not happen in nature except in the case of the mammalian placenta where the foetal and maternal tissues are in contact. The evolution of this organ had to overcome the antagonism of one protein system to another and the sort of difficulties that can arise are seen in the *Rhesus factor* of man.

The Rhesus factor can exist in two *allelomorphic* forms (i.e. genes situated at the same point on homologous chromosomes), a Rh^+ and a Rh^-, and these two forms are antagonistic to each other. Suppose the mother has Rh^- blood while the foetal blood is Rh^+ by parental inheritance, the Rh^+ being the dominant gene; then a small leakage occurs across the placenta and antibodies are generated in the maternal bloodstream against the foetal blood. As this antibody increases in each subsequent pregnancy the chances of damage to the foetus also increases, until a point is reached when its blood is clotted by antibodies passing into its circulation. The blood from such a child, one type of 'blue baby', must be replaced directly after birth if it is to survive.

A final type of antibody reaction takes place when living cells of skin, or of some organ, are grafted from one individual to another. Unless

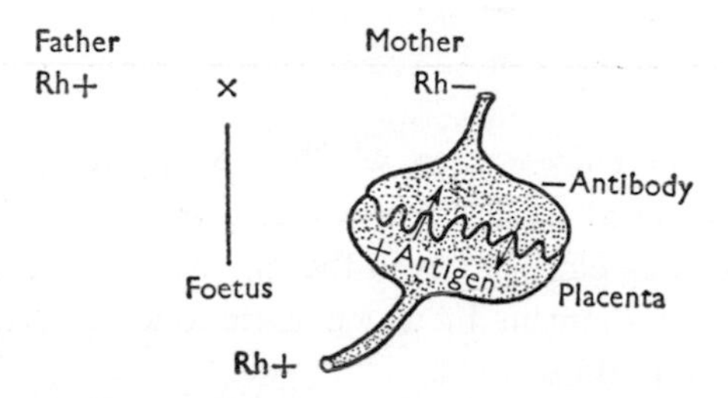

Fig. 63. The reaction between Rh^+ and Rh^- parent and offspring. The result is that antibody material from the mother enters the foetal circulation.

these individuals are *monozygotic* (i.e. identical) twins with the same gene complex there will be antibody reaction and the foreign protein will be destroyed. This represents a great barrier to surgery and methods of getting round this reaction, say by the application of cortisone, are only partly successful at present.

EXCRETION

The high degree of efficiency to which the physiological processes of mammals have evolved underlies the whole success and activity of the group. A part of this efficiency is in the maintenance of a stable internal (hence cellular) environment in which metabolic reactions can proceed under optimum conditions.

Two of the ways in which stability is achieved have already been described—the regulation of temperature and of the pH of the blood. No less exact is the control of the concentration and ionic composition of the body fluids. Metabolism means, essentially, the reactions between protein and other systems which make up all the activities of protoplasm. Protein reactions and properties depend not only on the pH and temperature but on the degree of hydration and the ionic medium in which they operate. Such important processes as contraction, conduction and membrane permeability are particularly dependent on the ionic medium; also, ions act in conjunction with many enzymes. The end-products of metabolic activities are often toxic to the organism. This is particularly true of nitrogenous substances from protein deamination, and the removal of these from the body is one of the main functions of the excretory organs.

The regulation of all these factors makes up the physiological function of osmotic and ionic regulation and excretion. Its main agents are the kidney, lungs and skin, and the substances involved are salts, water, carbon dioxide and nitrogenous waste. Although these organ systems have evolved independently for the mammal, efficient regulation as well as temperature control depend on their interaction.

A number of terms are used to describe the concentration of the body fluids and urine. It is important to know these in order to understand the following account of regulation. First there is the important concept of *osmosis*. This is the tendency of water to pass across a semi-permeable membrane to a medium of higher concentration. All living membranes are not truly semi-permeable because they allow the passage of ions and molecules other than water. Osmotic concentration is most effectively produced by small molecules or ions, so that the larger protein molecules of the body fluid do not contribute to its osmotic pressure in the same way as small salt molecules.

Isotonic is a term used to indicate that no water passes across a membrane separating two such solutes. Correspondingly a solution is *hypertonic* to another when water passes into it across a given membrane. The solution from which the water is lost is called *hypotonic*. These terms do not necessarily indicate the relative osmotic concentrations of the two solutions; these are given by the expressions *iso-osmotic*, *hyperosmotic* and *hypo-osmotic* respectively. In iso-osmotic solutions it would be possible, because of the properties of the membrane, for water to move from one to the other due to different tonicities.

6.1 Water balance

For most mammals water is an essential constituent of their diet—for a small number of desert dwellers it can be derived from the metabolism of foodstuffs. Whatever the source, the body of a mammal is some two-thirds water (man 63–67 %), and active protoplasm is always hydrated to this sort of proportion. Protoplasm is a colloid whose continuous phase is water, and as the cell membrane is permeable to water and some solutes, the concentration of the body fluids surrounding the cell must be similar to that of the cell sap if too much osmotic work is to be avoided. Water is used for the elimination of toxic nitrogenous waste, and although the urine of a mammal is usually hypertonic to the body fluids, salts cannot be concentrated beyond a certain point, so that salt excretion also needs water. Further use is in temperature control, where excess heat is lost by evaporation of water from the skin or lung.

Thus the demand over 24 hours can be as low as a litre or as high as 7 litres. If this is not met, the body becomes dehydrated and death occurs when some 20 % of the water content is lost.

Thus the demand over this period can be as low as a litre or as high as 7 litres. If this is not met, the body becomes dehydrated and death occurs when some 20 % of the water content is lost.

As mentioned above, the usual source of water is from the diet, either directly as a liquid or as a constituent of solid foodstuffs. Most vegetables are 80–90 % water and even dry foods, such as bread, contain as much as 10 % water. The water is taken up along the whole length of the alimentary canal, probably by the osmotic pull of the lining epithelium, unless the osmotic pressure of the gut contents happens to be higher than that of the epithelia, as it may be under certain conditions. The skin is almost impermeable to water, unlike that of the frog, so this is the only means of uptake.

Excess water taken in is rapidly eliminated from the body by the kidney. The mechanism of this regulation is described later.

Table 8

Organ	Water loss/24 hr (c.c.)
Kidney	600–2000
From gut in faeces	50–200
Skin	400–4000
Lung	350–400

6.2 Inorganic ions

Besides the specific uses of the various ions described below, much of the effective osmotic pressure of the cell and body fluid is due to the presence of these ions. Colloid osmotic pressure is very much less than that exerted by free ions. From its early evolution in the sea the cell has been surrounded by a chemical environment and it has needed to select certain ions, important in its functioning, and eliminate others. If the cell membrane is not to have a permanent charge the total sum of positively charged ions (cations) and negatively charged ions (anions) must be the same on each side. The cell has to use a number of devices to achieve a cytoplasmic environment which is isotonic with its surroundings. This need not be done by having the same concentration of osmotically active substances. Some of the devices for producing isotonicity are as follows:

(*a*) Differential permeability to various ions.

(*b*) The joining of an inorganic ion (e.g. K^+) with an oppositely charged organic ion which cannot pass the membrane (e.g. an amino acid), NH_3CH_2COO'.

(*c*) The active transport and removal of one ion in exchange for another.

It is the latter mechanism which is used a great deal in animals, and it is best known in the form of the sodium pump. This is a mechanism originally shown in the giant nerve fibres of the squid and is an active pumping out of Na^+ requiring metabolic energy. The removal of this cation from the cell allows other cations such as K^+ to enter—though ions may be concentrated within the cell by a 'pumping-in' mechanism.

This is found in nearly all mammalian cells as well as in other groups.

One can suppose that the early chemical environment in which the living organism arose was over-rich in Na^+ ions and that it was essential to acquire other ions inside—these could only be procured if a place was made for them and the ever-present Na^+ was an obvious ion to move out. On the other hand the sodium pump may have evolved in the cell as a means of reducing its osmotic problems—once again Na^+ being selected as the obvious ion to remove. Thus the cytoplasmic colloids became adapted to the other ions brought into the cell. As compared with sea-water the mammalian body and cell fluids are low in Na^+, Cl' and Mg^{++} (see Table 10).

Table 9. *The ions of the body fluids*

Ion	Where found	Use
K^+	Main cation of cell	(*a*) Contributes to osmotic pressure of cell (*b*) Provides potential across nerve cells (*c*) Activates enzymes (including phosphorylases)
Na^+	Main cation of body fluid	(*a*) Works with K^+ in producing potential across cell membrane (*b*) Contributes major osmotic pressure of body fluid
Ca^{++}	In body fluids mainly	(*a*) Affects permeability of cell, possibly involved in Na^+ pump action (it accumulates in over-stimulated nerve) (*b*) Involved in muscle contraction through activation of 'myosin' enzyme (*c*) Involved in the functioning of 'cement' holding cells together
Mg^{++}	Both in fluid and cell	(*a*) Decreases permeability of membranes but otherwise acts contra-osmotically to Ca^{++} (*b*) It decreases excitability of muscle (*c*) Has activating effect on some oxidising enzymes
PO_4''' HCO_3'	Both in body fluid and cell	(*a*) Use in buffering changes of pH
Cl'	Main anion of body fluids	(*a*) Affects permeability (*b*) Balances cation concentration within cell

The ions of the body fluids are taken in with the food and water of the diet. They are selectively absorbed by the cells of the alimentary canal but the main uptake is by diffusion across a gradient. Some have the specific uses outlined in the discussion on the diet (that is, as straightforward building materials—for example, in bone structure), but others have more complex roles and these are briefly summarised in Table 9.

The composition of seawater and of the urine and blood plasma of man is shown in Table 10.

In a very general sense the urine eliminates from the body the excess of the ions in which the composition of the body fluids most departs from seawater. However, the concentrations of these ions are by no means constant in the body even if some major differences from seawater can be seen.

Table 10

(In mM/kg)

	Na^+	K^+	Ca^{++}	Mg^{++}	Cl'	SO_4''
Plasma	143	5	5	2·2	103	1
Urine	143	35	6	9	136	20–60
Seawater	465	9·9	10·2	53	542	35

(It should be noted that seawater varies slightly in composition and urine very widely.)

A human red blood corpuscle, for example, contains Na^+ 10, K^+ 105, Ca^{++} 0, Mg^{++} 5·5, Cl' 80, SO_4'' 0—in other words the Na^+, Ca^{++} and Cl' are actively kept out while K^+ and Mg^{++} are concentrated. This is for specific reasons to do with the functioning of the corpuscle, but it shows what a large local variation in ionic composition the body can produce.

6.3 The kidney

The mammalian kidneys are made up of some 2,000,000 filtration units, called *nephrons*; the total length of these amounts to some 50 miles in the adult man. The kidney is a *metanephros* and has a long evolutionary history through the vertebrate line. The metanephros (found in *amniotes*[1]) is a compact organ working with the high hydrostatic pressure exerted by the blood. We are mainly concerned here with the mammals, but as some vestiges of earlier excretory systems are incorporated in the latter some account of the origin of the mammalian kidney should be given.

6.31 *Origin of the kidney*

The nephrons arise from the mesoderm surrounding the *nephrocoel* (or segmental excretory cavities of the embryo); there is one pair of *tubules* to each segment and they open into a duct which leads to the *cloaca*. Blood is supplied by the aorta to the nephrocoels and forms a capillary network—this is the *glomerulus*; the capillaries join up again and lead into the

[1] Those classes of vertebrates, namely the reptiles, birds and mammals, that have an amnion in their embryological development.

posterior cardinal veins. The tissue immediately surrounding the glomerulus is the *Bowman's capsule* and this, together with the glomerulus, makes up the *Malpighian body*. The whole excretory body plus its tubule forms the nephron.

This segmental plan is very primitive and does not exist in the adults of modern vertebrates. In mammals the excretory bodies and their tubules form in three areas—from the front of the animal to the back. These groups are called pro-, meso- and metanephros respectively—they are not separate organs but all arise from homologous embryological tissues.

The *pronephros* functions as the embryonic kidney but does not persist, though its ducts become incorporated with the collecting ducts of the mesonephric adult kidney. It is, however, found as the functional kidney of the tadpole.

The *mesonephros* is the functional kidney of fishes and amphibians. It loses its segmental arrangement as more tubules and Malpighian bodies develop. A part of it is found in the amniotes in the collecting ducts (or vas efferentia) from the testes.

The *metanephros* is only found in the reptiles, birds and mammals where it is the functional kidney: it has its own collecting duct—the *ureter*, and a great many extra excretory units—the nephrons, whose tubules often become very complicated.

The ureter leads down to a sac or *bladder* originally derived from the walls of the cloaca. This collects up the *urine* passed down from the kidneys and passes it out of the body from time to time via the *urethra*. Except in young mammals the sphincter muscle at the bladder is under conscious control, but emptying of the bladder is normally stimulated by the pressure of urine on proprioceptors within the bladder muscles.

6.32 How the kidney operates

In gross structure the mammalian kidney can be seen to consist of two main layers—an outer *cortex* and a central *medulla*. A variable portion of the medulla is made up of the pyramid of large collecting ducts which open into the ureter (Fig. 64). The nephrons (i.e. Malpighian body and tubules) are arranged with the filtration unit, the Malpighian body, in the cortex and a good part of the tubule in the medulla. The latter is much modified from the primitive anamniote condition and a *loop of Henle* is present between the proximal and distal convoluted tubules. This loop and the whole tubule is concerned with the complex process of salt concentration and reabsorption of useful substances filtered off from the blood. The descending and ascending limbs differ in their histological appearance in transverse section.

The fact that the two limbs of the loop of Henle run parallel to one another and to the collecting duct has recently been shown to be of physiological significance. The length of the loop is correlated with the size of the pyramids, which are most distinct in desert mammals and those that live in conditions of water shortage.

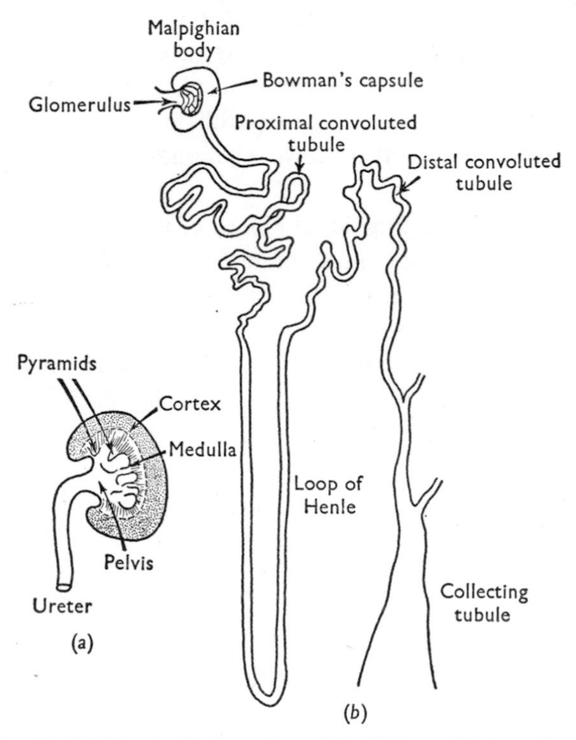

Fig. 64. (*a*) Longitudinal section of the kidney. (*b*) A nephron.

6.33 *Blood supply*

The blood supply of the tubule is very complicated (Fig. 65). A branch of the renal artery passes into the glomeruli—the combined effect of all the glomeruli is to filter from the blood about 130 c.c./min, representing about one-tenth of the blood passed through the kidney in this time, which itself is as much as a quarter of all the blood in the body. The capillaries of the glomeruli are particularly permeable (see below) and in all they make up 1–2 m^2 of surface-area across which the smaller molecules and ions can be passed.

The *renal artery* leaves the aorta and enters the kidney—here it splits up into *interlobar arteries* and these in turn into *circumferential arcuate arteries* (Fig. 65). The latter give off small *interlobular arteries* and from these the glomeruli branch off. The glomeruli have an afferent vessel, taking blood to them from the interlobular artery, and an efferent vessel collecting up the blood and passing it to *interlobular veins* and these in

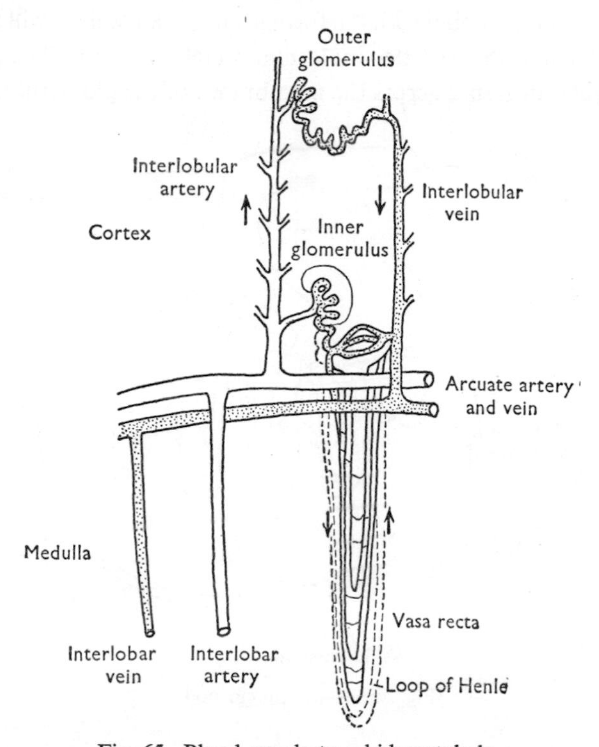

Fig. 65. Blood supply to a kidney tubule.

turn via *interlobar veins* to the main renal vein and thus back into the circulation.

Whereas the outlying glomeruli (which do not take part in osmoregulation) have the simple circulation described above, those lying near the medulla (or juxtamedullary) have quite different connections. The efferent vessel on leaving the glomerulus loops down in a series of U-shaped bends, called the *vasa recta*, and these come in close communication with the convoluted tubules and loop of Henle (Fig. 65).

6.34 *The glomerulus*

The walls of the capillaries which make up the glomerulus are permeable to substances of low molecular weight; these include water, nitrogenous waste such as urea, inorganic ions, and some useful substances such as sugar and occasionally protein. The size of molecule which cannot pass through the filter has a molecular weight which lies between 67,000 and 72,000. Thus haemoglobin with the former molecular weight will pass while serum albumen with the latter molecular weight will not. The passage of each filterable substance across the membranes of the glomerulus will vary

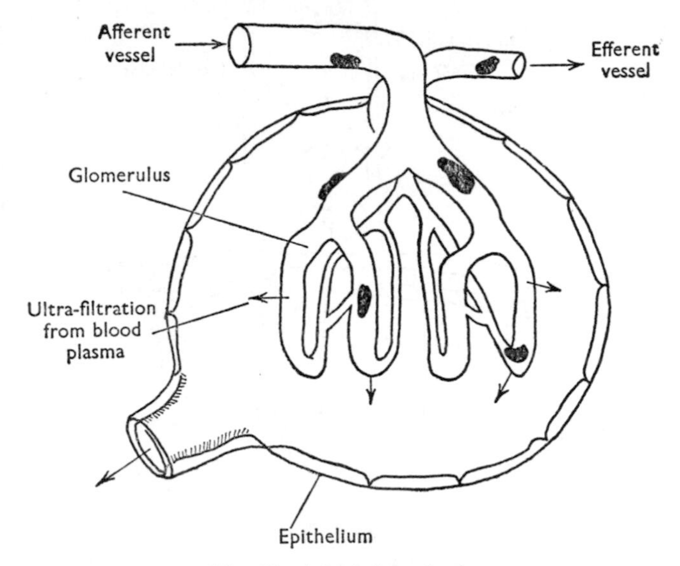

Fig. 66. A Malpigian body.

with its concentration in the blood, but by and large the composition of the filtrate for these smaller molecules and ions is the same as that of the blood plasma. However, by the time the urine reaches the bladder the amount of these substances is quite different, as shown in Table 11.

6.35 *The tubule and loop of Henle*

From the above figures it can be deduced that the tubule and loop of Henle are selectively removing most of the substances passed out of the glomerulus and that this activity can result in the formation of a *hypertonic urine*. In fact some 90–99 % of water as well as all useful substances and various amounts of salts are returned to the blood against the concentration gradient by means of an active process which uses large quantities of metabolic energy.

The mechanism of concentrating certain substances and eliminating others is still by no means clear but in recent years an explanation of the production of hypertonic urine and the role of the loop of Henle has been put forward.

Table 11

Substance	Amount passed in filtrate of glomerulus/day	Amount passed in urine/day
Na^+	600 g	6 g
K^+	35 g	2 g
Ca^{++}	5 g	0·2 g
Glucose	200 g	Trace
Water	180 litres	1·5 litres
Urea	60 g	35 g

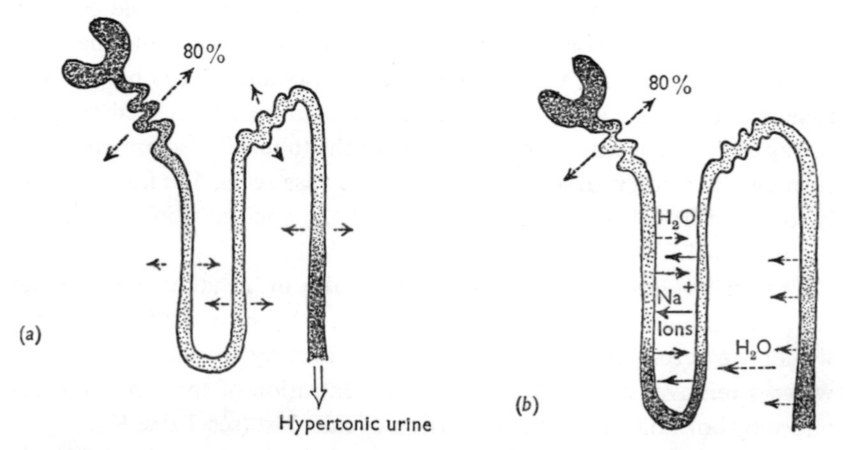

Fig. 67. (*a*) Concentrations in nephron (old view). Gradual removal of water along the tubule. (*b*) Concentrations in nephron (modern data). Passage of ions and uptake of water by counter-current hypothesis.

Early work on the frog, which has a large Malpighian body, showed the general function of the parts of the nephron. In the mammal it is known that some 80 % of the salts and water passed out of the capsule is re-absorbed in the proximal convoluted tubule (Fig. 67 (*a*)). It was assumed that the production of a hypertonic urine depended on the gradual uptake of water along the length of the remainder of the tubule and by the loop of Henle. This would have suggested a concentration gradient along the tubule as shown on Fig. 67 (*a*). Measurements of the ionic concentrations of portions of kidney made in the last few years have shown that there is an increase of concentration from the cortex to the medulla, so that, within each tubule and loop of Henle the picture is as illustrated in Fig. 67 (*b*). In other words there is not a gradual change in concentration along the length but an increase towards the loop, a decrease towards the distal tubule and a final concentration in the collecting duct and pyramid.

The explanation of these observations is found in the *Counter-current Multiplier Hypothesis* suggested in 1954. This hypothesis suggests that as the glomerular filtrate passes up the ascending limb of the loop of Henle Na^+ is actively absorbed, and transported across to the descending limb into which it is actively returned. By this means the concentration of the filtrate in the loop is continuously built up until it becomes hypertonic to the body fluids. Exactly how this transport of the Na^+ takes place is not important here, nor indeed is it fully understood, but the idea of a build-up of concentration by the multiplication of these small differences of ion concentration between the two limbs of the loop of Henle should be clear.

The filtrate now passes up the ascending limb of the loop into the distal convoluted tubule and, with the removal of Na^+ ions, it has become iso-tonic to the body fluid. At this stage it should be appreciated that the strong osmotic concentration in the loop of the tubule is also present in the surrounding medium and in the vessels of the vasa recta. The flow of blood is so slow through these latter vessels that the concentration gradient is not dispersed.

The final stage in the concentration of the urine takes place in the collecting tubule. If the body needs to make a hypertonic urine, water is withdrawn across the walls of the collecting duct by osmosis. How much water is removed and thus the actual concentration of the urine is deter-mined by hormones as is the proportion of salts lost (see Table 12).

From this explanation it can be understood why the region of Henle takes the form of a loop and why the vessels of the vasa recta take the form of loops. Increase of the limbs of these loops leads to more efficient concentration of the glomerular filtrate and in desert mammals the loop of Henle is very long. Such mammals can produce urine that is more than twice the concentration of seawater, a remarkable physiological feat, and most mammals are able to produce urine somewhat more concentrated than seawater. The human kidney has relatively short loops of Henle and is not able to make urine as strong as this, which is why man cannot use salt water as his only source of liquid.

The ability to excrete hypertonic urine is essential for successful ter-restrial life and is found well developed in reptiles, birds, and insects.

For the human kidney the ability to concentrate excretory substances including salts, when this is necessary, is shown in Table 12:

Table 12

Substance	Ratio: body fluid/urine
Urea	1: 60
Water	1: 1
Creatinine	1:100
Na^+	1: 1
K^+	1: 7
SO_4''	1: 60
NH_3	1:400

From the above it is obvious that the kidney is a highly effective excretory and osmo-regulatory organ.

Besides these two functions the kidney can also assist in the control of body pH (complementing the buffer action of the blood). This is kept strictly between 7·35 and 7·45 and the role of the kidney in maintaining this consists of excretion of acid radicles such as HPO_4' and reabsorption of alkali ions (Na^+). Ammonia is also passed into the tubule in varying amounts allowing a more acidic urine to be excreted without upsetting Na^+, K^+ and other cation reabsorption.

6.36 *Control of the kidney*

The blood flow to the kidney as well as the activity of the tubule is variable. Increase in blood pressure, say by adrenalin, will increase the quantity of glomerular filtrate formed. Cooling of the body can depress take-up of water by the tubules. Both these effects lead to increase in urine flow.

Conversely any factor depressing the blood pressure, such as haemorrhage, will cause less filtrate to be formed through the glomerulus and thus lower urine production.

The actual regulation of Na and water content of the urine are under the control of two hormones. The *anti-diuretic hormone* (ADH) from the *posterior lobe* of the *pituitary* controls the permeability of the end cells of the tubule and by increasing this more water is taken up from the filtrate as it passes along this region. Thus the amount of water leaving the body can be controlled within fairly wide limits although there is a definite maximum concentration possible (1 water:1·022 salt). The ADH is released by the pituitary in response to the increased osmotic pressure of the blood in the hypothalamus. The ADH is released from the pituitary after having been produced in the neurosecretory cells (see p. 230) of the hypothalamus and migrating down the axons of those cells to the pituitary.

Sodium reabsorption across the walls of the loop of Henle is controlled by an adrenal cortex hormone whose release is itself governed by a pineal hormone, adreno-glomerulo-trophin. Again the operation of this mineral corticoid hormone controls elimination of salt (or its retention) by the

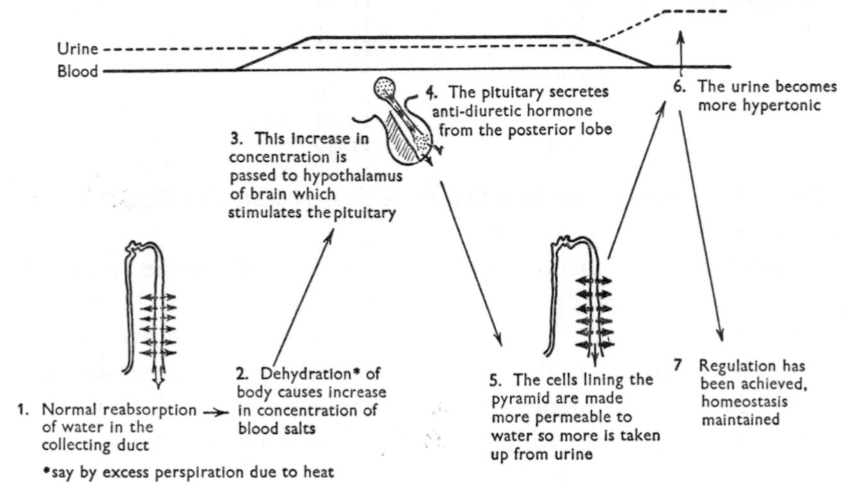

Fig. 68. Endocrine regulation of urine concentration.

body, and again there is a maximum and minimum rate of its loss. The maximum loss is less than 3 g/day. As seawater contains many times this amount/2000 c.c. it can be seen that a man who was forced to take this volume/day to maintain his water balance would accumulate so much salt that he would eventually die.

6.4 Other organs of excretion: the skin and lung, bile, faeces

The skin and lung have been dealt with on p. 80 and p. 49.

The gall bladder and bile ducts also act as true excretory organs and the bile contains two breakdown products of haemoglobin—*bilirubin* and *biliverdin*. This system has been described on p. 33.

Finally the faecal matter contains some 40 % by weight of the bodies of bacteria from the large intestine, some 15 % of indigestible substances such as cellulose, 45 % water and is coloured by bile pigments. Faeces is not really an excretory product, with the exception of the bile substances, as its contents have never been incorporated into the body.

6.5 The sources of nitrogenous excretory substances in the urine of mammals

6.51 *Urea*

Urea, $CO(NH_2)_2$, is formed from the deamination of amino acids taken in with the diet. It is thus derived from exogenous protein originating outside the body. Urea is formed in a cyclic process called the *ornithine cycle.* Carbon dioxide and ammonia combine with the amino acid ornithine to form the amino acid citrulline which in turn takes up a further molecule of ammonia to give arginine. Hydrolysis of this by the enzyme arginase yields a molecule of urea and a new molecule of ornithine which recommences the cycle (see p. 273). Urea is found in mammalian blood at some 30 mg/100 ml and is the main nitrogenous excretory substance.

6.52 *Ammonia*

Ammonia, NH_3, is very toxic and is not released into the blood in any quantity in mammals (conc. approx. 0·02 mg/100 ml). It is made by deamination of amino acids in the liver where it is rapidly converted to urea by the route described above. It is also made in the kidney, whence it enters the urine in small quantities.

6.53 *Creatinine*

A complex nitrogen-containing substance derived from the breakdown of tissue (hence *endogenous*) protein.

6.6 A comparative account of water and salt regulation in the vertebrates

(The osmotic pressure of the body fluids of animals are customarily measured by the depression of the freezing point they cause. This is expressed by the symbol Δ, the internal medium being represented by Δ_i and the external medium as Δ_o.)

6.61 *Fishes*

6.611 *Freshwater fishes.* Fishes arose in freshwater and the present seawater species of both elasmobranchs and teleosts have such an ancestry. In freshwater the surrounding medium is low in salts and the internal osmotic pressure of the fishes' body fluids are very much higher than that

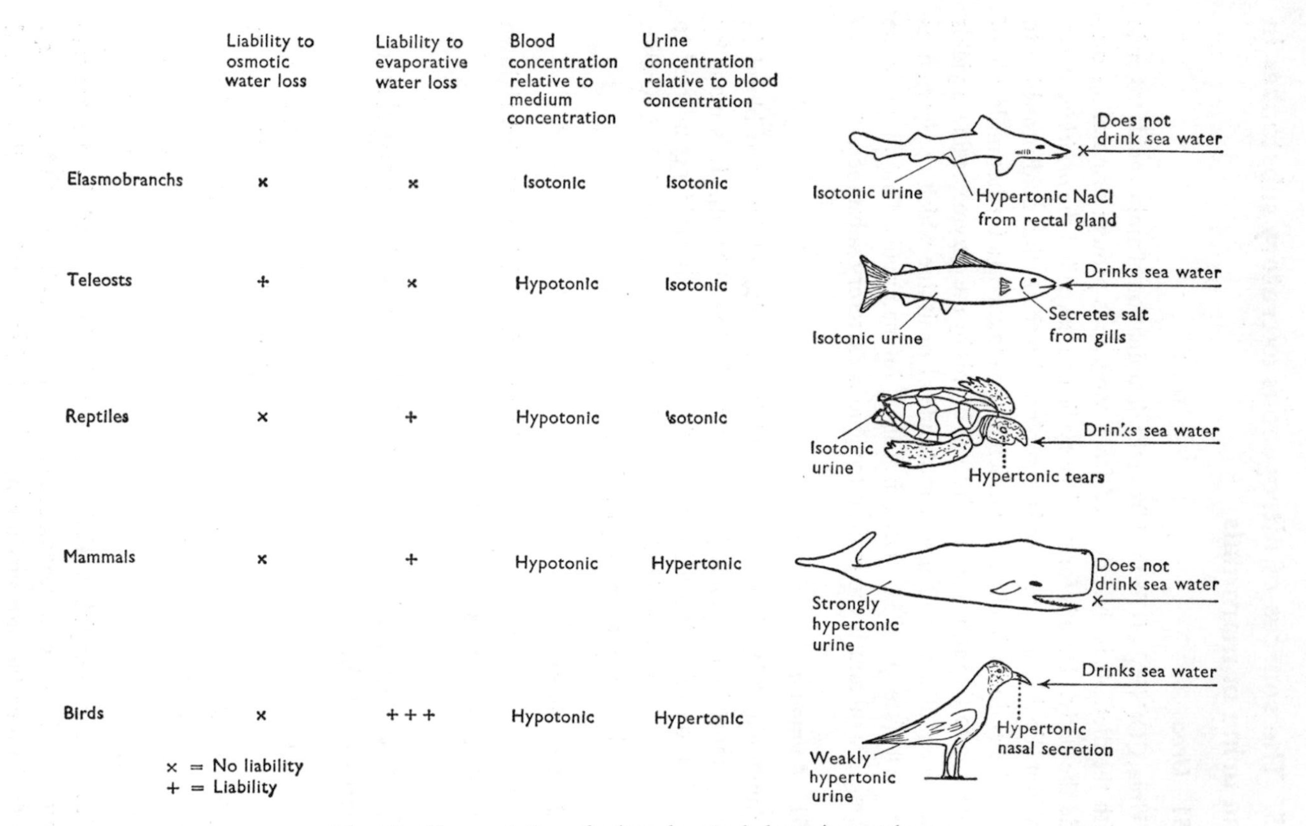

	Liability to osmotic water loss	Liability to evaporative water loss	Blood concentration relative to medium concentration	Urine concentration relative to blood concentration
Elasmobranchs	×	×	Isotonic	Isotonic
Teleosts	+	×	Hypotonic	Isotonic
Reptiles	×	+	Hypotonic	Isotonic
Mammals	×	+	Hypotonic	Hypertonic
Birds	×	+ + +	Hypotonic	Hypertonic

× = No liability
+ = Liability

Fig. 69. The regulation of salt and water balance in vertebrates.

of the environment ($\Delta_i = 0{\cdot}57$, $\Delta_o = 0{\cdot}03$). Water will thus enter the body through permeable surfaces such as the gills and gut but the skin itself[1] is relatively impermeable. In freshwater fishes large quantities of water are filtered by the glomeruli into the tubules where most of the salts are re-absorbed. Because salts are lost in the urine, which is hypertonic to the medium, they must be made good in the diet. There are also special salt-absorbing cells on the gills. Freshwater fishes have large Malpighian capsules associated with the passage of considerable quantities of water. Freshwater elasmobranchs which have invaded this medium from the sea still retain urea in their bloodstream (see below).

6.612 *Marine fishes.* (*a*) *Teleosts.* Seawater has an osmotic pressure greater than that of the blood of marine teleosts, whose body fluids ($\Delta_i = 0{\cdot}6$) tend to have a similar composition to those of their freshwater relatives. The osmotic pressure of the sea corresponds to $\Delta_o = 1{\cdot}7$. Marine teleosts thus lose water through the gills, gut or any permeable surfaces, and their osmotic and ionic need is to dilute the external medium. In order to do this the fishes swallow large quantities of salt water, absorbing both the water and the salts from the gut. They conserve the water, producing only small quantities of urine, and eliminate excess salts via special secretory cells on the gills as well as in the urine. Some of the excretion of nitrogenous waste takes place through the gut lining. Because of the small amount of water eliminated the capsule and glomerulus are very small.

(*b*) *Elasmobranchs* that live in the sea are unusual in using the retention of urea in the body as a means of raising the internal osmotic pressure.[2] In these fishes the $\Delta_i \simeq 1{\cdot}8$ and that of seawater $\Delta_o \simeq 1{\cdot}7$. For this reason the marine elasmobranch has no tendency to lose water to the medium and its kidney has a large capsule and well-developed tubule for selective reabsorption of salts (see also p. 46). The young elasmobranch is provided with a supply of urea from its mother.

6.62 Amphibians

The amphibians evolved from a group of freshwater fishes in the Devonian (a geological period that commenced some 300 million years ago) and they have many of the same problems of osmo-regulation and ionic maintenance. The amphibians live largely in freshwater and have a

[1] The mucus which covers the eel during its time in freshwater is thought to be a means of decreasing the permeability of its skin.

[2] A similar adaptation is found in some marine frogs.

body fluid which is hypertonic to the surrounding medium (for *Rana*, the common frog, $\Delta_i = 0{\cdot}56$). The skin is partly permeable to water as well as the gut lining and thus amphibians absorb large quantities of water from their freshwater environment. Although the kidney has a large glomerulus and capsule as well as a long tubule for salt reabsorption some leaching out of salts occurs. The urine is copious and hypotonic to the body fluids ($\Delta_o = 0{\cdot}17$). The loss of salts is made up in the diet as well as by selective absorption through the skin.

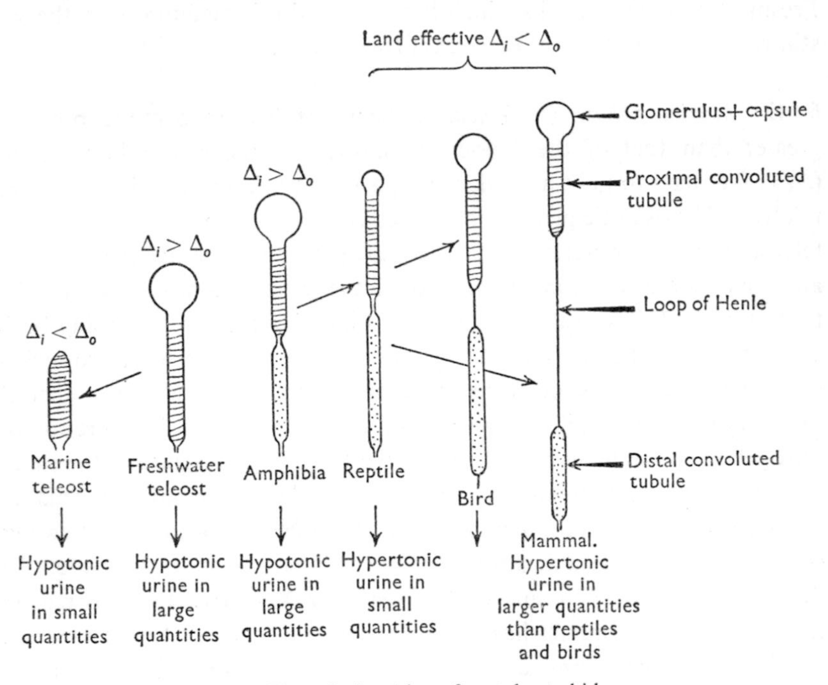

Fig. 70. The relationships of vertebrate kidneys.

6.63 *Reptiles*

Reptiles, like all land animals, have a serious problem of desiccation by their environment however (unlike amphibians) their scaly skin is nearly impermeable. The supply of water available is limited and some water must be lost in the elimination of toxic waste products. Water is taken in with the diet and there is only a small glomerulus and capsule so that a low filtration rate into the tubule takes place. The urine produced is concentrated ($\Delta_o \simeq 0{\cdot}6$) and may be semi-solid. The ability to produce such concentrated urine lies in the reptiles' use of the fairly insoluble and non-toxic uric acid as the main nitrogenous excretory product. This substance requires less water than ammonia or urea for its elimination.

Some reptiles have colonised the sea, for example, the turtle, and these have similar osmotic adaptations to the sea-birds described below.

6.64 *Birds*

The birds are closely related to reptiles and have the same problems of osmo-regulation and excretion. Unlike the former they have a large glomerulus and capsule and rely on the uptake from the long tubule to return water and salts to the bloodstream. A new element is present in the tubule of birds and is also present in mammals; this is called the loop of Henle and is an integral part of salt and water uptake in these two classes. Its development in the birds, however, is much less than in the mammals.

Water is also reabsorbed from the cloaca of birds, through which the urine passes, and like the reptiles they make use of uric acid, producing a semi-solid and hypertonic urine. In the large eggs of both reptiles and birds the nitrogenous waste products are stored up as the insoluble, non-toxic *allantoic acid.* The evolution of some such substance was essential for the metabolism of the shelled egg.

Marine birds and marine reptiles have an internal osmotic pressure which is less than that of their environment so that their problems are much the same as those of marine teleosts. Whether they drink sea-water or feed on animals whose body fluids are isotonic with seawater, marine reptiles and birds have an excessive intake of salts. The surplus salt is eliminated from a secretory gland on the head called the *nasal gland.* In marine birds this gland is some 10 × to 100 × larger than its homologue in terrestrial forms. The nasal gland exudes a salt solution more concentrated than sea-water and thus eliminates any excess salts taken into the body, which is the reason for such birds and turtles shedding tears.

6.65 *Mammals*

The mammals have already been dealt with above. They can secrete a hypertonic urine and have a large glomerulus and capsule with a long tubule and loop of Henle for reabsorption of water and salts in the tubule.

THE SKELETON, MUSCLES AND MOVEMENT

7.1 Bone and the skeleton

7.11 *Function*

Bone forms the main skeletal element of the body. It provides not only support but also protection, and gives a jointed structure of struts and levers which is operated by the muscles. In the red *marrow* of the bone the red and certain of the white corpuscles are made.

7.12 *Structure*

In the adult mammal, bone consists of numerous *trabeculae* or lines of bone-forming cells and bone matrix oriented with respect to the forces acting on the bone. Both spongy and compact bone is full of the vessel-carrying Haversian canals from which the *osteoblasts* (bone-forming cells) derive food and oxygen. These osteoblasts lie in lacunae or spaces, arranged concentrically from a central canal. Each osteoblast produces a fine canal system so that there is no part of the bone matrix far from a strand of living protoplasm (Fig. 71).

The matrix itself is a rigid substance composed of some two-thirds inorganic calcium phosphate and the rest of an organic fibrous protein similar to collagen. The properties of bone in its ability to resist tension and compression compare very favourably with granite or concrete. During its growth the long bones of mammals ossify from three centres. The main shaft or *diaphysis* is separated at each end from an *epiphysis*. The epiphyses fuse with the diaphysis when the bone no longer grows.

7.13 *The joints*

Movement takes place where the bones meet at the joints. Their nature varies according to the amount of movement that they permit. In the skull the bones are joined firmly across sutures by the *periosteum*, the connective tissue which covers all bones. Where there is relatively little movement, as between vertebrae, the joints are pads of cartilage, but movement is much easier at *synovial* joints.

Fig. 71. Structure of bone in transverse section.

A synovial joint, such as that between a limb and the limb girdle, consists of two articulating members capped with *hyaline* (transparent) cartilage and enclosed by a *capsular ligament*. To the outside of the cavity of the joint lie the synovial membranes whose function is to secrete the lubricating synovial fluid. This fluid is also derived from the cartilage itself and gives a sort of hydraulic suspension to the body. The outer layers of the capsule have fibrous thickening which hold the bones together.

7.14 *Adaptations of bone*

Bone is a tissue designed to withstand tension and compression. It must be strong enough to take the standing weight of the body and the forces exerted by that weight in moving. These forces may be very great—for example, a ten stone man, whilst running, may exert some 3500 kg/cm² on the shaft of the femur.

The forces acting on bone are met in ways similar to those used by engineers in the construction of girders, cranes, bridges, etc. It can be seen

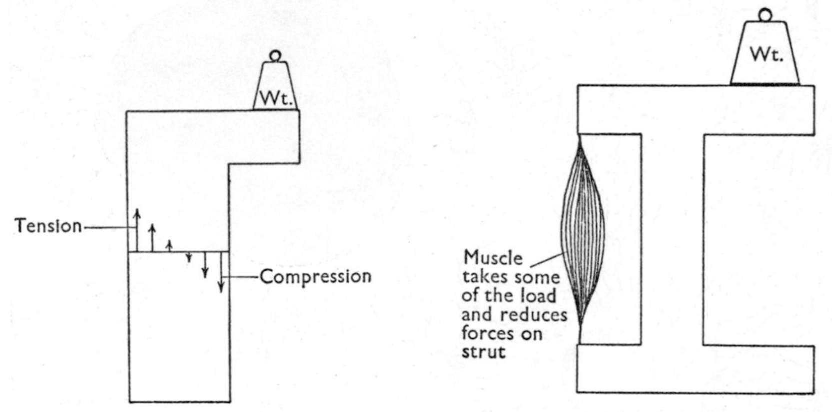

Fig. 72. Forces carried by a solid strut loaded eccentrically.

Fig. 73. The reduction of tension on a bone by use of muscle connexions.

(Fig. 72) that forces applied to solid struts eccentrically produce compression and tension in the outer rim of the strut. A solid strut can thus be replaced by a hollow one, which has the effect of lightening the skeletal material without reducing its strength—especially found in birds.

Another device serving to reduce the forces acting on a bone is the use of braces of muscle or tendon, such as the biceps which runs from shoulder to forearm, or the ilio-tibial tract along the outside of the thigh. These act as a bow-string and balance some of the forces of tension acting on the far side with respect to the weight, and also reduce to a lesser extent compression forces on the opposite side (Fig. 73).

7.2 The skeleton of the mammal

It is necessary, before considering how the skeleton becomes modified for various means of locomotion, to have a clear idea of the position and function of the various bones. The arrangement of bones in a common terrestrial mammal, such as the rabbit, represents the results of a long evolution and adaptation towards an efficient performance in terms of support, protection, and especially locomotion on land.

It is useful to consider the skeleton of a mammal to be made up of two major parts: the *axial skeleton*, which includes the skull, ribs and vertebral column, while the *appendicular* (or hanging) *skeleton* includes the limb girdles and limb bones.

7.21 *The axial skeleton*

7.211 *The skull.* The names of the bones that make up the skull will not be described here—sufficient to say that they make a well-protected box in which the brain can be housed and that the mandible, or lower jaw, operates efficiently according to the feeding habits of the animal (see p. 42). The skull articulates with the first neck vertebra by two large occipital condyles.

7.212 *The vertebral column.* Around the embryonic *notochord* (a skeletal rod) ossify a number of skeletal blocks or vertebrae. These fall into five sections in the mammal and the numbers in each section, except for the last, are constant. Working down from the skull we find 7 cervical or neck vertebrae, 12 thoracic or chest vertebrae, 7 lumbar or back vertebrae, and a sacrum consisting of 4 fused vertebrae. The caudal, or tail, vertebrae are variable. The vertebrae play an important role in all functions of the skeleton, they protect the spinal cord which runs along inside them, they support the weight of the head, tail and abdomen, and they act as the origin for powerful back muscles essential in locomotion.

(*a*) *The cervical vertebrae* (Figs. 74–76). These are characterised first by a large neural canal, the spinal cord having just left the medulla; secondly, by well-developed centra and small projecting processes, because they act as compression members (the weight of the head is taken by ligaments and muscles attached to the neural spines of the thoracic vertebrae) and need to be flexible. The first two cervical vertebrae are modified to form a universal joint for the head. The skull articulates by its occipital condyles with the concave facets of the first cervical or *atlas* vertebra (Fig. 74). This allows the head to move backwards and forwards (or in the rabbit—up and down) across the large flat joint that is formed. The atlas also has a wing-like transverse process for the origin of the muscles which move the head in this plane. It has no centrum.

Behind the atlas the second cervical vertebra, the *axis* (Fig. 75), allows the rotation of the head. It is narrow and its centrum extends as a peg, the odontoid process, which projects into the atlas. In fact this process was derived from the centrum of the latter.

RABBIT BONES

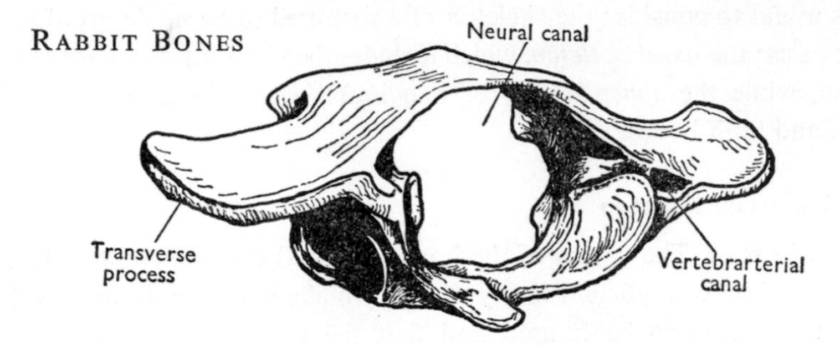

Fig. 74. Atlas vertebra.

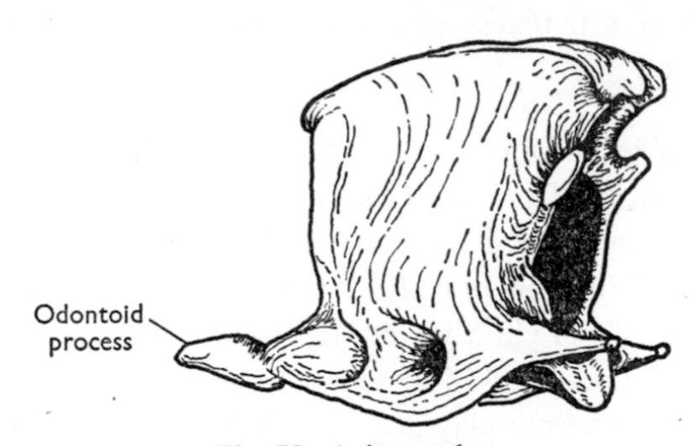

Fig. 75. Axis vertebra.

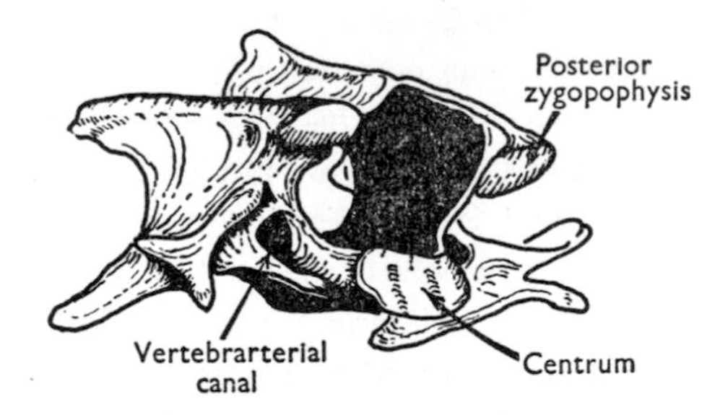

Fig. 76. Cervical vertebra.

The other five cervical vertebrae (Fig. 76) have large neural canals, reduced spines and transverse processes and large vertebrarterial canals at each side. The *zygapophyses* of the individual vertebrae are flattened, the *prezygapophysis* pointing dorsally and the *postzygapophysis* ventrally. This increases flexibility of the neck.

(*b*) *The thoracic vertebrae* (Fig. 77). The long, backwardly directed neural spines of these vertebrae are characteristic and are well seen in the rabbit. They vary in length according to the size of the head and carry the muscles and *ligamentum nuchae* which act, in this region, as the tension members of the vertebral column, being attached from the neural spines of the thoracic vertebrae to the neck. The vertebrae themselves are compression members of the supporting system so formed and their centra are

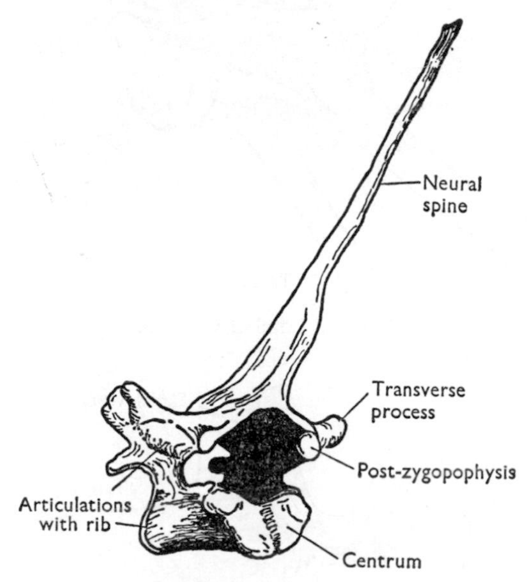

Fig. 77. Thoracic vertebra.

larger than in the cervical vertebrae and the neural canals are smaller as nerves pass out from the spinal cord. The transverse processes are reduced to a broad facet with which the tubercle of the rib articulates.

(*c*) *The lumbar vertebrae* (Fig. 78). These may be easily recognised by the long, anteriorly directed transverse processes which provide areas of attachment for the large back muscles (the *sacrospinalis* and *longissimus dorsi*) involved in bending the back and in transferring the weight carried by the posterior part of the vertebral column to the pelvic girdle. The neural spines, which are short, are also anteriorly directed and provide further points of origin for the back muscles. (Thus between the thoracic and lumbar vertebrae the stresses are coming from opposite directions and the facets and spines of the two sections therefore face in reverse directions to meet these forces. The junction is at the diaphragmatic vertebra.)

Associated with the more powerful muscles of the pelvic region the centra of the lumbar vertebrae are better developed than those of the previous type to take the compression forces involved in locomotion.

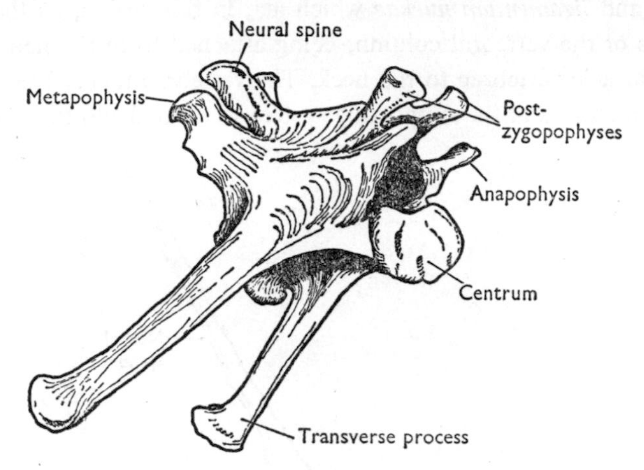

Fig. 78. Lumbar vertebra.

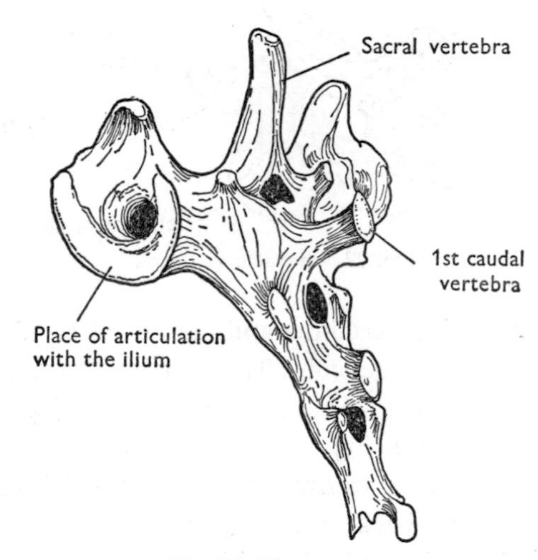

Fig. 79. The sacrum.

(*d*) *The sacrum* (Fig. 79). The function of the sacrum is to provide a firm attachment of the vertebral column to the pelvic girdle. Unlike the pectoral girdle, which acts partly as a shock absorber, the main force of locomotion from the muscles of the hind limb and buttocks passes through the pelvic girdle–sacral articulation, which allows very little movement.

The sacrum consists of four fused vertebrae of which the first is modified with large articular surfaces to fit against a similar surface on the inside of each ilium. The three posterior members of the sacrum are small and lead directly into the caudal vertebrae.

(*e*) *Caudal vertebrae.* The rabbit has a short tail and the 15 caudal vertebrae which support the tail are small cylindrical bones. Their articulation is such as to allow a freer movement between them than other parts of the vertebral column.

7.213 *The Ribs.* These long curved bones are inserted by their heads or *capitula* against the centra of the thoracic vertebrae, while a small dorsal spine, or *tuberculum*, supports them against the transverse processes of the same vertebrae. The ventral portion of the rib is of cartilage and is attached to a rectangular sternum for ribs 1–7 but not for 8–12 of which the last three are called floating ribs. The degree to which the ribs move depends on their importance in the support of the body via the pectoral girdle. In animals whose weight is taken on four legs the movement of the ribs in ventilation of the lungs is less important than that of the diaphragm.

7.22 The appendicular skeleton

7.221 *The fore limb.* (*a*) *The pectoral girdle* is the means of transmitting the weight of the front part of the animal's body to the fore limb. It consists of a large flattened *scapula*, with a spine running down the centre, and a very small *clavicle* articulating with the base of the scapula. The scapula is attached to the vertebral column by four sets of muscles. Inside muscles run to the rib cage, outside they run from the thoracic vertebrae, anteriorly to the anterior thoracic vertebrae and posteriorly to the posterior thoracic vertebrae. Thus the pectoral girdle is firmly anchored to the axial skeleton, but unlike the pelvic, the anchorage is made by muscles and there is no direct skeletal connection (see Fig. 80). This allows more spring in the suspension of the front half of the body, which is particularly important during leaping or running when the animal takes the shock of landing on its front rather than its hind limbs. From both the inside and outside of the scapula, muscles take their origin and serve to restrict movement of the fore limb to a single anterior-posterior plane.

At the base of the scapula is a small spine, the metacromion, which is directed posteriorly for the articulation of the clavicle, while below it the large saucer-shaped glenoid cavity provides a seating for the head of the humerus. Anteriorly a *coracoid process* is used for some muscle insertions.

(*b*) *The humerus* is a stout bone with the proximal end enlarged into a head (articulating with the glenoid of the scapula). On either side of the head, processes called *tuberosities* are the sites of insertion of muscles such as the deltoid and pectoralis and others from the scapula.

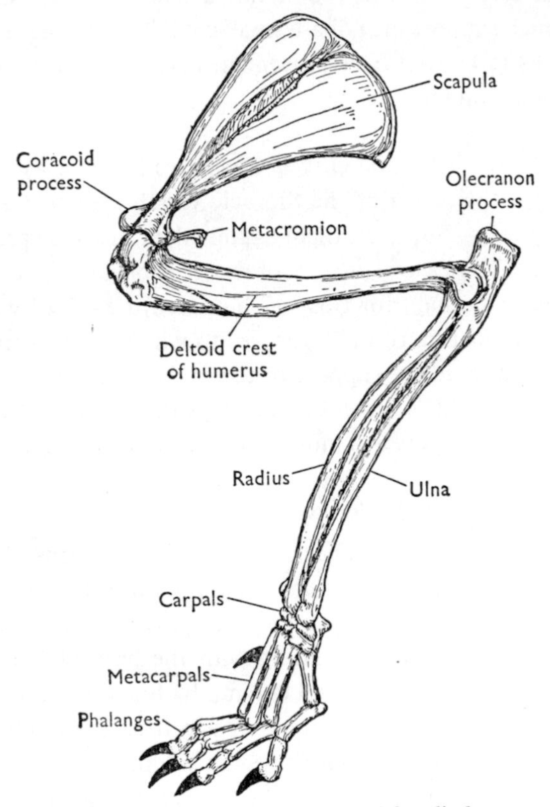

Fig. 80. Left pectoral girdle and fore limb.

At the further, or distal end of the humerus, is the condyle for insertion into the sigmoid notch of the ulna. This forms a characteristic hinge-joint which limits movement to a single plane.

(*c*) *Radius and ulna* are two long curved bones making up the shank of the limb. In the rabbit the radius is anterior and shorter than the ulna and is fixed to each end of it. The proximal end of the radius makes up the lower part of the sigmoid notch into which the humerus fits, and the radius takes most of the weight, which it transfers to the foot via a distal articulation. The ulna is the main bone for the attachment of muscles in this segment of the limb. Its proximal end extends out beyond the sigmoid notch into the *olecranon process*, and on this the triceps, the main extensor of the fore limb, is inserted. The distal end of the ulna articulates, like the radius, with some small bones at the top of the foot.

(*d*) *The fore foot*. There are many of these bones or *carpals* at the proximal end of the foot. The number of these is very variable in mammals. In the rabbit there are two rows, the first containing the radiale, the intermedium and the ulnare, and the second containing the trapezium, the trapezoid, the centrale, the magnum and the unciform. A small extra bone called the pisiform is found on the outer carpals of mammals (and reptiles). All these bones give a flexible joint (the wrist) between the arm and the hand.

Passing down from the carpals are the elongated *metacarpals*. Again these are very variable in different mammals but in the rabbit fore limb there are five. Each consists of a long bone attached at one end to the carpals and at the other to the digits. Between these two sets of bones the ends of the metacarpals make hinge joints for extension or flexion of the digits or *phalanges*.

The digits are long cylindrical bones which end in claws. There are five in the rabbit fore limb and this is taken to be the ancestral number—hence the description of the tetrapod limb as *pentadactyl*.

7.222 *The hind limb*. (*a*) *The pelvic girdle* is made up of three bones on each side of the body. The *ilium* is the largest and has a flattened wing running forwards. The inside of this is articulated by ligaments to the 1st sacral vertebra so that the joint is very strong and nearly fixed. From the outer surface of the ilium originate the large retractor muscles (see p. 169) of the hind limb.

Behind the ilium the *ischium* extends posteriorly and this again acts as a point of origin of the retractors of the limb. The latter are very powerful muscles of great importance in locomotion. The third bone, the *pubis*, lies inside and in front of the other two and forms a bony cage around the genitals. Between this and the ischium is the large obturator foramen, and to the outside where the three bones meet there is a large cavity—the *acetabulum*, for articulation of the femur.

(*b*) *The femur*, like the humerus, is a stout cylindrical bone with enlarged ends for articulation. At the proximal end there is a round head protruding to the inside and fitting into the socket of the acetabulum. Along the shaft of the bone next to the head there are a number of wing-like processes, the trochanters, for the attachment of muscles. There is also a crest running down the dorsal length of the bone for similar purposes. At its distal end the femur has two condyles for the articulation with the tibia. Once again there are wing-like surfaces for muscle attachments and a characteristic feature is the deep groove dorsally for the tendon of the extensor muscle of the lower part of the limb.

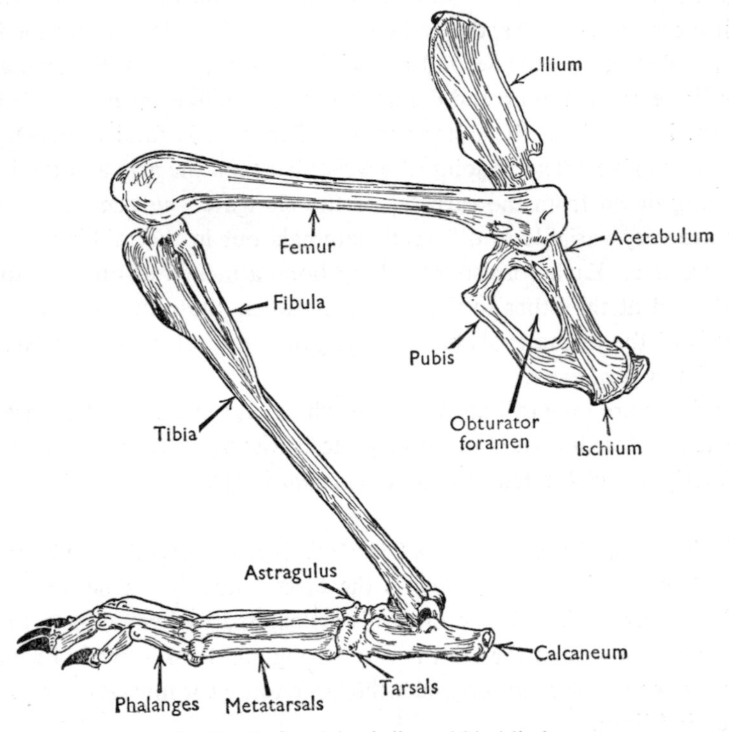

Fig. 81. Left pelvic girdle and hind limb.

(*c*) *The tibia and fibula.* The first of these bones is large and carries all the weight down to the hind feet. It has a flattened head which articulates with the condyle of the femur forming a hinge joint and at the side of the head there is the *cnemial* (lower leg) crest for attachment of the *patella* or knee-cap. On to the latter are inserted the tendons for the extension of the tibia. The distal end of the tibia articulates with the tarsus of the hind foot. In the rabbit, as in many mammals, the fibula has become vestigial and is fused to the distal end of the tibia.

(*d*) *The hind foot.* The ankle corresponds to the wrist and is made up of small bones—the *tarsals.* The first row consists of two substantial elements, the astragalus and calcaneum, with which the end of the tibia articulates.

The calcaneum is prolonged posteriorly to provide a point of insertion for the calf muscle which extends the foot and gives a degree of mechanical advantage to this muscle which is important in running and jumping.

Below the astragalus is the square centrale and this is followed by a second row of tarsals—the mesocuneiform, ectocuneiform and cutoid. As with the wrist these bones of the ankle joint are little modified from the primitive pentadactyl limb.

Following the tarsals are four elongated *metatarsals* which make up the main shaft of the foot. The first digits have been lost. Finally there are the four phalanges, which end in claws.

7.3 Cartilage and connective tissue

7.31 *Cartilage*

7.311 *Function.* Cartilage makes up the embryonic skeleton but remains only at the ends of bones in the adult, the rest becoming ossified. It is also found in isolated places in the body such as the external ear, epiglottis, etc. In every case it is a supporting tissue well able to resist compression and intermediate in tensile strength between connective tissue and bone. Where cartilage persists as a main skeletal element in the adult (e.g. dogfish) it is often strengthened by deposition of calcium salts.

7.312 *Structure.* The living cells of cartilage are *chondroblasts*; they arise from a surrounding *perichondrium* and secrete a ground mass of chondrin. The properties of cartilage are due to this semi-liquid matrix of *chondrin* in its containing membrane. Hyaline cartilage has only the above substances, whereas fibrous and elastic cartilage have collagen and elastin in their make-up. There is no direct blood supply to the chondroblasts so that their food and oxygen must diffuse across the matrix (Fig. 82).

7.32 Connective tissue

7.321 *Function.* These tissues support and hold together other cells and tissue systems of the body. They also provide part of the defence mechanism. Connective tissues are distributed under the skin, around the nerves and vessels, between muscle and bone, bone and bone, as lubricating mesenteries, synovial membranes and as fatty tissue surrounding the kidney and other organs. They also form the periosteum covering the surface of bone, providing a means of attachment for tendons and ligaments and giving rise to new bone.

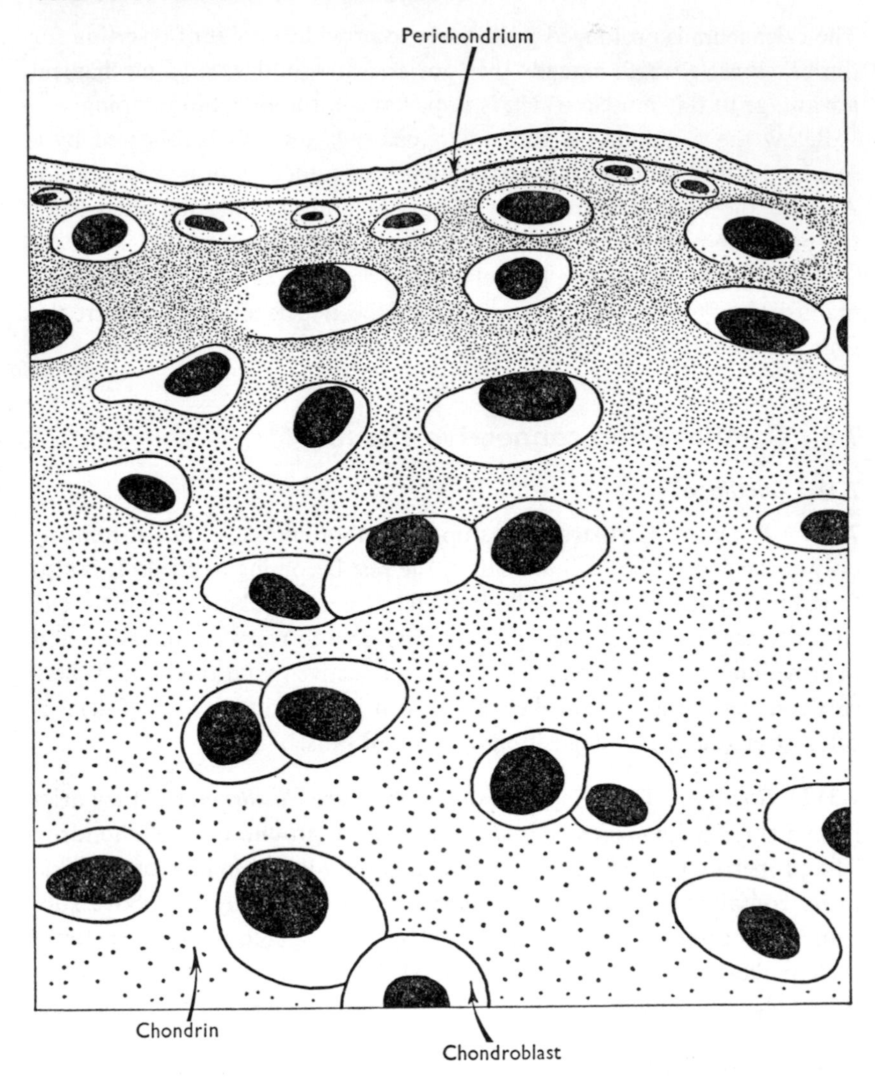

Fig. 82. Cartilage.

7.322 *Structure.* The basic cells are fibroblasts which are able to secrete the other structures found within the tissue. In all types of connective tissue a matrix of mucopolysaccharide is secreted in which the fibroblasts lie scattered, while further substances laid down depend on the position and function of the tissue. White fibres of collagen are commonly found where resistance to stretching is required. The collagen consists of protein, formed as unbranched fibres, up to 10μ in thickness. It makes up the bulk of tendons, ligaments and the general areolar tissue between organs. Yellow fibres of elastin, also secreted by the fibroblasts, are found in the connective tissue between vertebrae and around vessels. Elastin is unlike collagen in that it will stretch freely and subsequently return to its original size. It is present in areolar tissue.

The connective tissue also plays a part in the defence of the body and a form of white corpuscle called a macrophage is found scattered throughout the tissues. These are similar in appearance to the fibroblasts, from which they probably originate, but have the property of ingesting bacteria and can secrete antibodies. Together with similar cells of the lymph, spleen and liver these white corpuscles make up the reticulo-endothelial system.

7.4 Muscles

7.41 *Function*

Muscles perform the work which moves parts of the skeleton relative to one another; they cannot actively extend but are pulled out usually by the action of another or *antagonistic* muscle. In the alimentary canal there are longitudinal and circular groups of muscle which again are antagonistic to one another and operate against the hydrostatic skeleton formed by the gut contents. Muscles can generally be divided into plain and striated. Cardiac muscle is striated, it is also a *syncytium.* Plain muscles are under autonomic or involuntary control and are particularly adapted to produce changes in length and operate in visceral activities. Striated skeletal muscle is under direct central control (voluntary) and is capable of changes in tension but relatively small changes in length. Striated muscles are used during locomotion, respiration, etc.

7.42 *Structure*

A skeletal muscle is made up of many striated fibres, each of which may be quite long and may exceed 10 cm in length. They are usually between 10 and 100μ in diameter and altogether they make up more than 50 % of the weight of the body. At each end of the muscle the fibres are united by blunt ends to tendons which are usually attached to bones but sometimes to connective tissue. The finer tapering ends of the fibres end in connective tissue within the muscle itself. Each muscle fibre contains hundreds of nuclei and is enclosed in a connective tissue membrane—the *sarcolemma.* Internally the fibre is composed of many *myofibrils* of about 2μ diameter and the remaining protoplasm forms the sarcoplasm (Fig. 83). Some

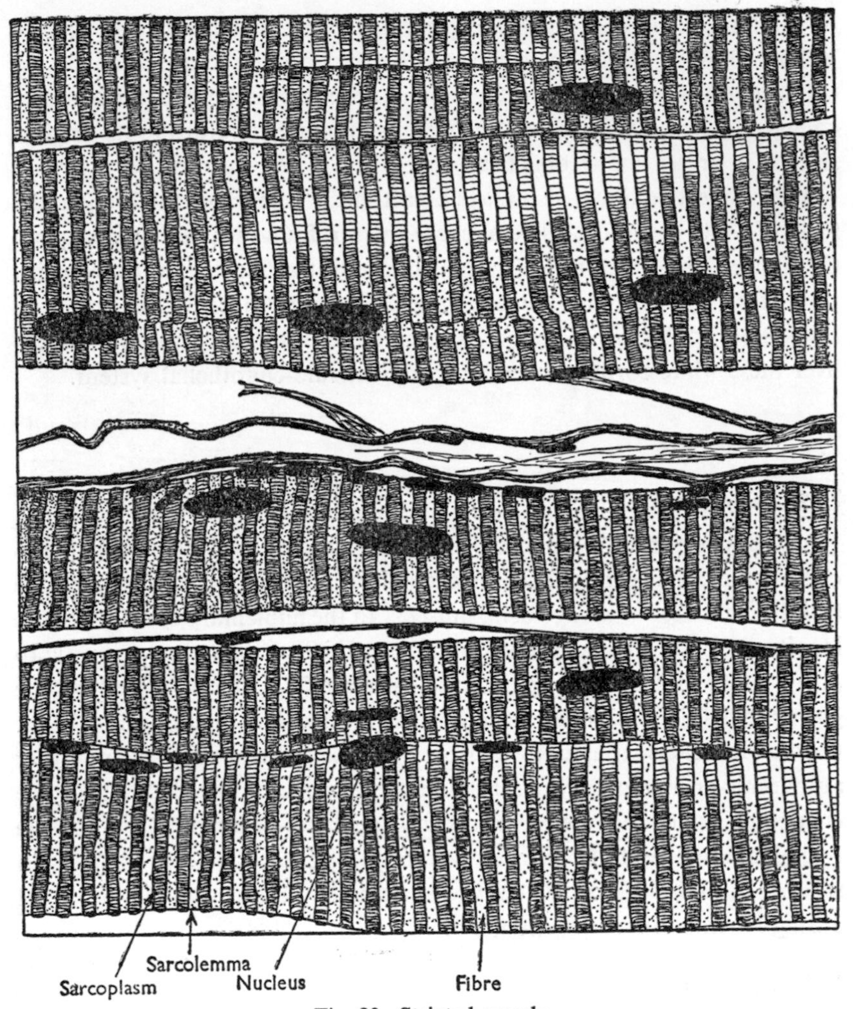

Fig. 83. Striated muscle.

muscles are red in colour because of the large content of myoglobin and these are generally concerned with sustained contractions. Other white muscles are concerned with the more rapid or phasic movements of the organism.

7.43 *Innervation*

Each muscle is supplied by many nerve fibres, each of which innervates a number of muscle fibres (Fig. 84). When a nerve impulse passes down a nerve fibre it propagates in an all-or-none way and excites all of the muscle fibres which it innervates. In this way the muscle is made up of numerous *motor units* each of which can function as a whole, and the total activity of the muscle is made up of the addition of the individual activities of these units. Normal skeletal muscles have 2000–3000 fibres in each motor unit but the extrinsic eye muscles have only 10–15 and hence their contractions can be graded in smaller quanta. Where the nerve fibres terminate on the muscle fibres there is a single *end plate*. It is at this point that the nerve

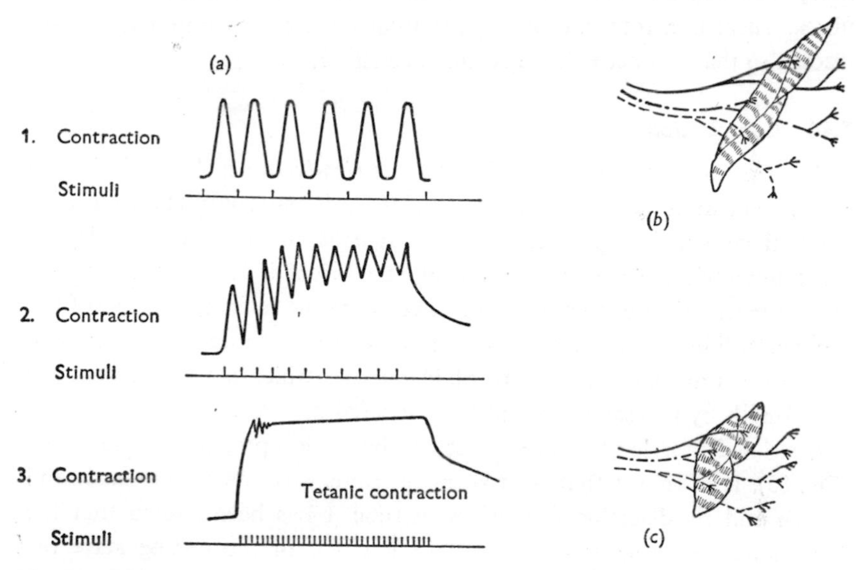

Fig. 84. Diagram showing the behaviour of a single muscle fibre in each of three motor units. (*a*) The nature of muscle contraction depends on the frequency of the stimulus. Tetanic contraction occurs when successive stimuli arrive before the muscle has had time to relax. The maximum contraction is developed by a muscle when all the motor neurones leading to it are stimulated (*c*). Less powerful contractions result when a limited proportion of fibres are working (*b*).

impulse liberates a small amount of chemical, acetylcholine, which changes the permeability of the muscle membrane, leading to the propagation of an action potential along the muscle fibre (Fig. 103). Contraction of the fibre is coupled to the action potential in a way which is not yet clear. After such a sequence of events has occurred, the muscle fibre takes a short time to re-establish the charge on its surface (as happens in nerve fibres) and during this refractory period the muscle membrane cannot be excited. Excitation can occur, however, before the mechanical contraction has declined and when this happens at the arrival of another impulse there is a summation of the mechanical contractions which results in a *tetanus* (Fig. 84).

The strength of a muscular contraction is determined by the number of motor units that are active at any one time. Under normal conditions it is rare for all of them to be active at once, although this condition (the maximal twitch) is often used in physiological experiments. When a mammal is in its normal posture and at rest the tension exerted by a muscle remains fairly constant, but this is not due to the contraction of the same muscle fibres. There is a rotation among the total muscle-fibre population contracted so that the net tension remains constant.

7.44 Contraction

A given muscle may contract either *isometrically*, in which case its length remains constant but it increases its tension, or it may contract *isotonically*, where there is a change in length but no tension change; for example, the more isometric the contraction the more efficient it is because the contractile mechanism does not have to overcome the internal friction of the muscle. In life most muscles shorten to a greater or lesser extent. Isotonic contractions are more typical of plain muscles which are able to change their length by a greater amount than striated muscles.

A striated muscle fibre can be seen to be made up of transverse bands (Fig. 85). It is known that the two main proteins of muscle are *actin* and *myosin* and by digesting out one at a time it has been shown that the distribution of these proteins is as in Fig. 85. In the resting state the respiratory product adenosine triphosphate (ATP), which contains energy-rich phosphate bonds (see p. 265), is attached to the myosin. Contraction takes place when this ATP is caused to break down and as it does so some of the energy released catalyses the combination of actin and myosin proteins. The remainder of the energy released is set free as heat and the ATP breaks down to the diphosphate (ADP). The actomyosin formed is a complex protein shorter in length than its separate components and the actin becomes drawn into the A bands of the myosin. During contraction it can be seen that the I bands decrease in length and the A bands do not, which supports the above supposition.

After contraction has ceased—that is, when the nerve impulses are no longer arriving at the end plate and the original stimulus for contraction is withdrawn—ATP once more recombines with the myosin. At the same time the actin-myosin link is broken and this allows the muscle to relax—but of course it needs to be pulled back into its original shape. During

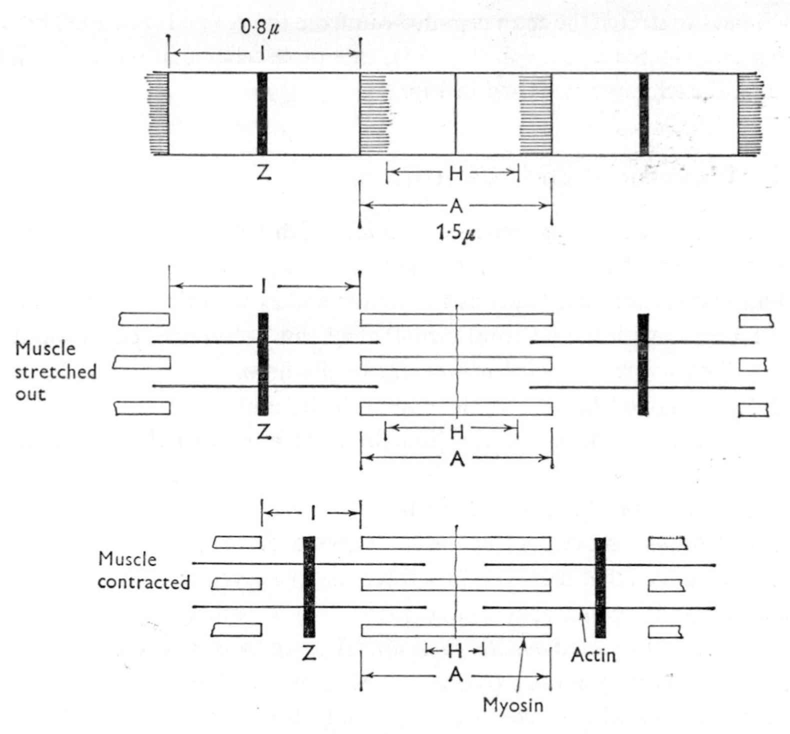

Fig. 85. Striated muscle, showing the changes that take place on contraction.

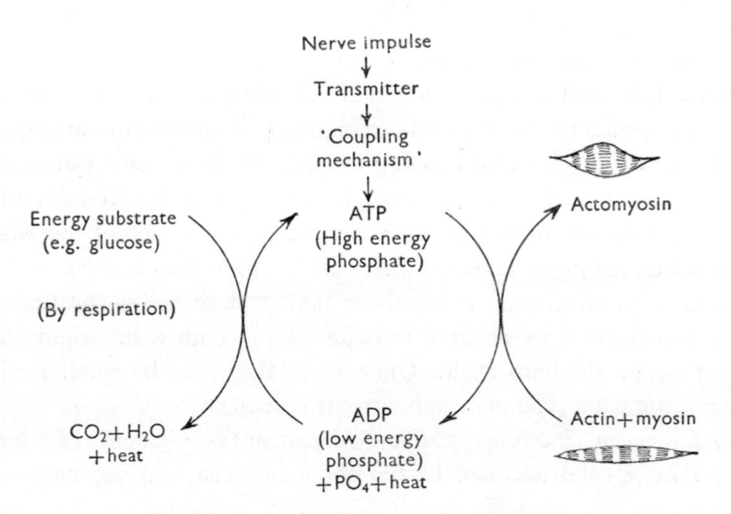

Fig. 86. Role of ATP in muscle contraction.

such passive stretch the actin is pulled out from the A bands and the I bands are seen to increase in width (Fig. 85). A simple scheme of the energy and chemical exchanges is shown in Fig. 86.

7.5 Movement and the muscles

The muscles working across a limb joint can be classified according to the movements they produce. Considering the possible movements about a ball-and-socket joint (such as the femur makes with the pelvic girdle or the humerus with the pectoral girdle) they can be subdivided as follows:

(*a*) *Protraction*—forward movement of the limb.

(*b*) *Retraction*—backward movement of the limb.

(*c*) *Elevation*—raising of the limb at right angles to the longitudinal body axis.

(*d*) *Depression*—lowering of the limb in the same plane.

(*e*) *Rotation*—movement of the limb about its longitudinal axis.

It can be seen that these fall into three major movements which can take place in opposite directions so that (*a*) and (*b*), (*c*) and (*d*), and (*e*) may be brought about by *antagonistic muscles*. The action of many limb muscles combines several of these movements. Muscles which produce movements of the limb as a whole have their origin outside the limb although they are inserted on one of the limb bones, usually the most proximal (femur or humerus). Muscles with their origin outside the limb are called *extrinsic* limb muscles, but it must be understood that this classification is entirely a functional one and has no implications about the homologies of these muscles. The relative importance of the different extrinsic muscles varies from one species to another and consequently their size and arrangement will vary. In walking and running animals the propulsive power comes from retraction of the limbs pushing the feet against the ground and thus driving the animal forwards. In birds the power stroke of the wings is delivered by the large depressor muscles (the pectoralis major).

Other sorts of movement are those taking place within the limb itself and are performed by *intrinsic* muscles having both their origin and insertion within the limb itself. Once again they may be subdivided into antagonistic pairs. The main movements consist of:

(*a*) *Extension*—increasing the angle between two segments of a limb.

(*b*) *Flexion*—the decrease in the angle between two segments of the limb.

These terms can be applied to the muscle systems operating across any joint. In a fairly unmodified form they may be seen in the fore limb of a frog (Fig. 87). The hind limb is modified with elongated ankle bones and powerful extensor muscles for leaping, but the fore limb is still fairly close to that of the ancestral type found among the first terrestrial vertebrates.

7.51 *Fore limb of the frog*

The propulsive force is produced by contraction of the retractor muscles, behind the joint: the *latissimus dorsi* above and the *posterior pectoralis* and *coraco-brachialis* below. Protraction, the forward motion of the limb, is due to the action of the *deltoideus*. Elevation and depression of the limb are brought about by muscles which also function as protractors and retractors. Important depressor muscles are the *median and anterior pectoralis* and elevation is mainly carried out by the *dorsalis scapulae*. Rotation of the humerus takes place when it is retracted by the posterior pectoralis and this is antagonised by rotation in the opposite direction when the anterior and median pectoralis and the dorsalis scapulae contract. Extension at the elbow joint is produced by the *triceps* but, unlike the human, a frog has no *biceps* to antagonise this action. Flexion of the elbow is produced by the *supracoracoideus* muscle which runs from the coracoid to the radius and ulna. This muscle is an example of a type, quite frequently found among vertebrates, in which the muscle operates across two joints; an arrangement which reduces the total amount of work done by the muscles of the body during locomotion.

The description of the extrinsic and intrinsic muscles of the frog's fore limb should be followed by dissection in conjunction with Fig. 87. By inserting a seeker into the back of the frog's skull and moving it about, the brain may be destroyed. This is called pithing and may be made more extensive by inserting the seeker into the vertebral column and so destroying the spinal cord. In this way the animal is rendered completely insensitive and yet the muscles retain their excitability. If the frog is now skinned and the muscles stimulated by electrodes—one being inserted into the body and the other applied to various muscles, the systems described can be readily understood by observing the movements of the limb. If this technique is not possible, however, satisfactory results can be obtained by dissection of the various muscles from their point of origin or insertion and pulling them with forceps to observe resulting movements.

It can be seen, therefore, that the various sets of muscles all bring about movements by contracting across joints. As muscle fibres can only shorten by one fifth of their length, the body has to compromise in the leverage produced by the bones between mechanical efficiency and high-velocity ratio. On the whole it uses a short force arm and a long load arm (as in the elbow), which means that the limited distance that the muscle can contract is magnified by the lever into a wide arc and the limb tip therefore

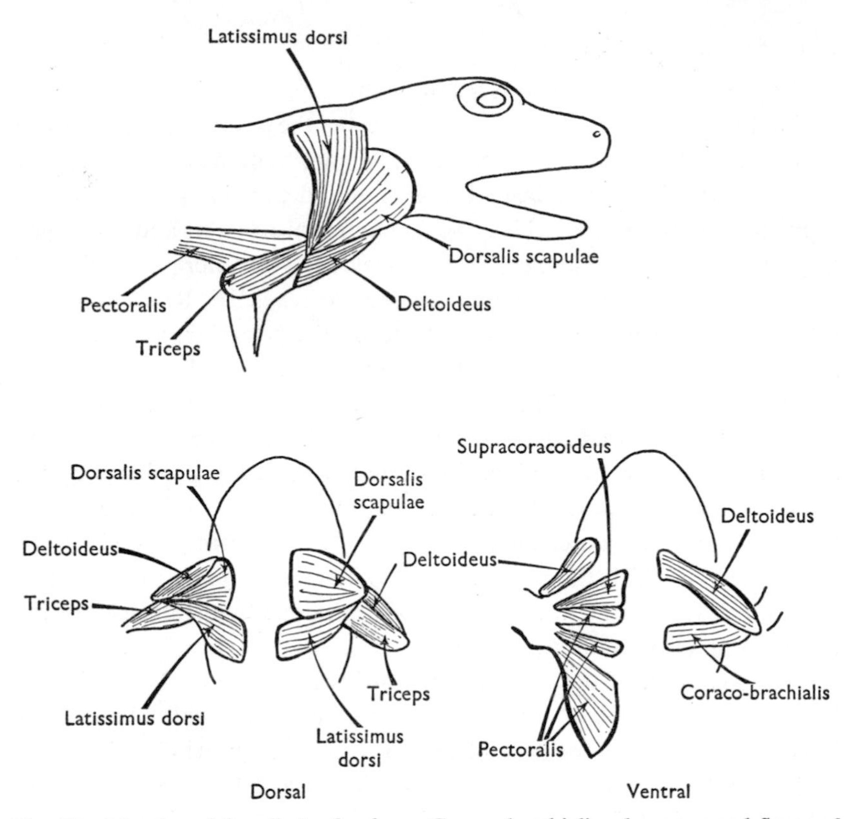

Fig. 87. Muscles of fore limb of a frog. Coraco-brachialis: depressor and flexor of humerus; deltoideus: protractor and elevator of humerus; dorsalis scapulae: protractor and elevator of humerus; latissimus dorsi: elevator and retractor of humerus; pectoralis: retractor and depressor of humerus; supracoracoideus: depressor of humerus; triceps: elevator of humerus and extensor of ulna.

moves at a relatively high velocity (Fig. 88). Such an arrangement is found in the retractor muscles of the horse's leg where high velocity of the hoof is required, but in digging animals the distance of the insertion from the fulcrum is greater and this enables them to exert a relatively larger force at the tip of the limb although it can only move slower. In addition to producing movements about joints, other muscles play an important role in keeping the bones in position at a joint.

If the joint considered in Fig. 88 is the elbow, then the flexor muscle is the biceps. Flexion of the elbow is also produced by the brachioradialis muscle, which has its origin on the humerus nearer the elbow and is inserted on the lower arm near the wrist, and remains more or less parallel to the forearm throughout its range of movement. Contraction of the biceps

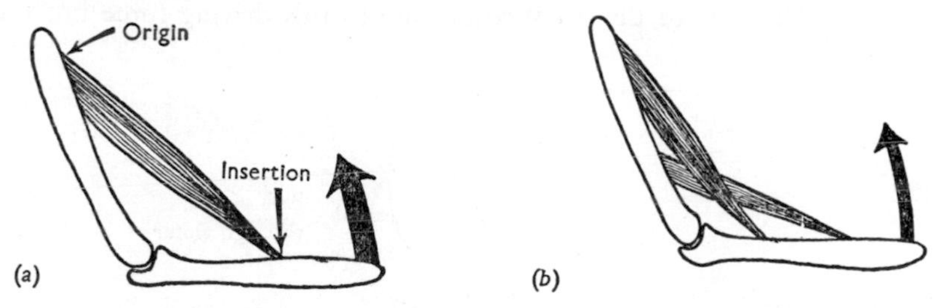

Fig. 88. Diagram showing how change in relative positions of origin and insertion of muscles alters their effective pull.

accelerates the lower limb mainly along the curve of motion and tends to be particularly active during slow flexion and the maintenance of a fixed position against gravity or a load. The brachioradialis, on the other hand, is only active during rapid movements when it serves to stabilise the elbow joint. It does this during rapid extension as well as flexion.

7.6 Types of locomotion used by vertebrates and adaptations of the skeletal and muscular systems

7.61 *Swimming*

This is the form of locomotion found among primitive vertebrates and is best developed in fishes but is also found as a secondary adaptation in other groups such as whales and dolphins. The main adaptations for movement in water include:

(*a*) *Streamlined shape*—this may involve reduction of the appendicular skeleton in secondarily adapted forms; for example, whales have a minute pelvic girdle.

(*b*) *Well-developed muscles* along the axial skeleton (*myotomes*) for producing undulatory movements, and usually a form of tail fin. In whales and dolphins the movements are in a dorso-ventral plane, whereas in fishes, including flatfishes, they are lateral. The 'tail fins' of mammalian swimmers are referred to as the flukes.

(*c*) *Stabilising elements.* Among fishes the paired and unpaired fins function this way but they may also play a part in active locomotion. This is also true among groups of secondarily aquatic vertebrates including seals, penguins and turtles, which make use of their paired limbs as paddles.

When a fish such as a dogfish is swimming, lateral waves of contraction are seen to pass backwards along the body. In a simple way these may be regarded as pushing against the water and hence driving the animal forwards. All parts of the body contribute to this driving force but the

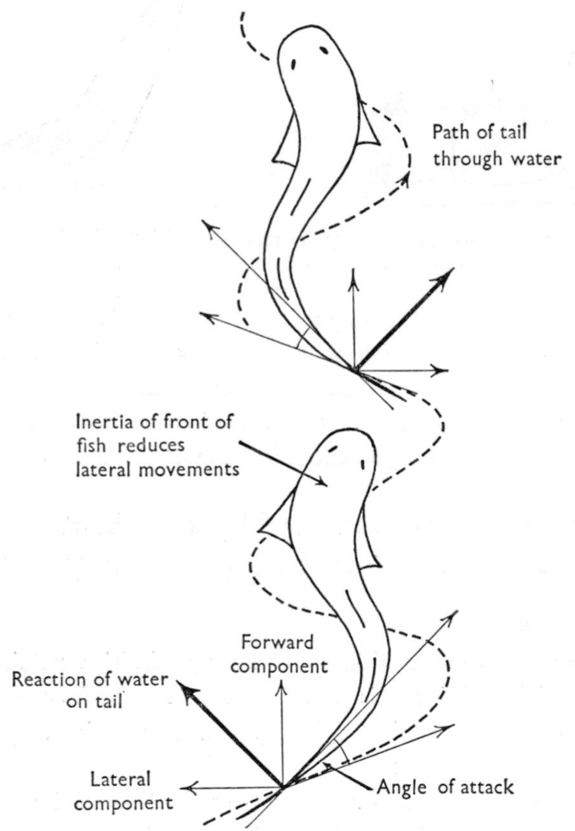

Fig. 89. Forces developed by the tail of a fish during swimming.

expanded tail probably provides a dominant share. From cine films it is clear that each segment of the body, and the tail, passes through the water at an angle to its surface. In this way they press against the water and the latter reacts on the body and produces a resultant force at right angles to the body surface. This force can be resolved into two component forces. A lateral one pushes the body sideways and is counteracted by equal and opposite forces acting at other parts of the body. The forward component is the force which overcomes the resistance of the water. In most bony fishes (eels are notable exceptions) the backward passage of waves along the body is not so clear. In these cases the most obvious movement is a sideways movement of the tail fin. This again acts as an inclined plane passing through the water at an angle to its path of motion and from it forces are developed. In these fishes the head scarcely moves to the side at all, whereas in the dogfish and eels the head is displaced laterally during the movements. This lack of lateral displacement of the head in most teleost fishes is due to the greater inertia at the front end and the resistance to sideways movement provided by its shape and the unpaired fins. In mammals such as a dolphin the propulsive thrust develops almost entirely from the flukes which once more are moved at an angle to their direction of motion in an up-and-downward plane. In all these cases of undulatory propulsion the speed is determined by the frequency of tail beat and especially by the length of the animal. A rough generalisation is that the maximum speed of a small to medium-sized fish is about 10 times its body length per second. A trout 1 ft long, for instance, can swim at 10 ft/sec. The highest speed recorded for a fish is about 40 ft/sec, or 27 m.p.h., for a 4 ft barracuda. Authentic speeds for dolphins are in the region of 15–20 knots. The ability of these animals to move so rapidly is aided by the properties of their skins which reduce their resistance considerably. It may even be true that the flow of water past a swimming dolphin is *laminar* (i.e. without any formation of eddies).

The muscles which produce these movements in fishes are the myotomes, with their characteristic W-shape when viewed from the side. This pattern is the outward sign of many segmentally arranged cones of muscle which interlock with one another. The myotomes operate between sheets of connective tissue (*myocommata*) which connect with the transverse processes of the vertebral column. This complex arrangement of the muscle fibres probably enables a smoother passage of the wave of contraction along the body and also prevents the formation of excessive thickenings of the body muscles in the region where they have contracted during the passage of the wave. Each segmental myotome contracts slightly ahead of the myotome behind and the waves pass alternately down each side of the body. The waves are coordinated by *proprioceptive* information which enters the dorsal roots. If all the dorsal roots of a dogfish are severed the animal cannot swim (p. 209).

The cigar-shape of aquatic animals, though favouring motion through the water, tends to roll about its longitudinal axis and to yaw and pitch

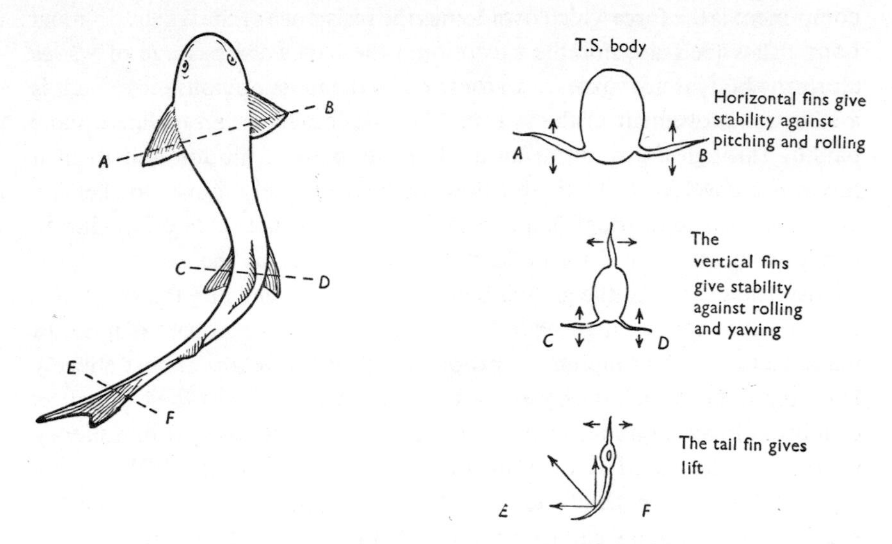

Fig. 90. Role of the fins in a swimming dogfish.

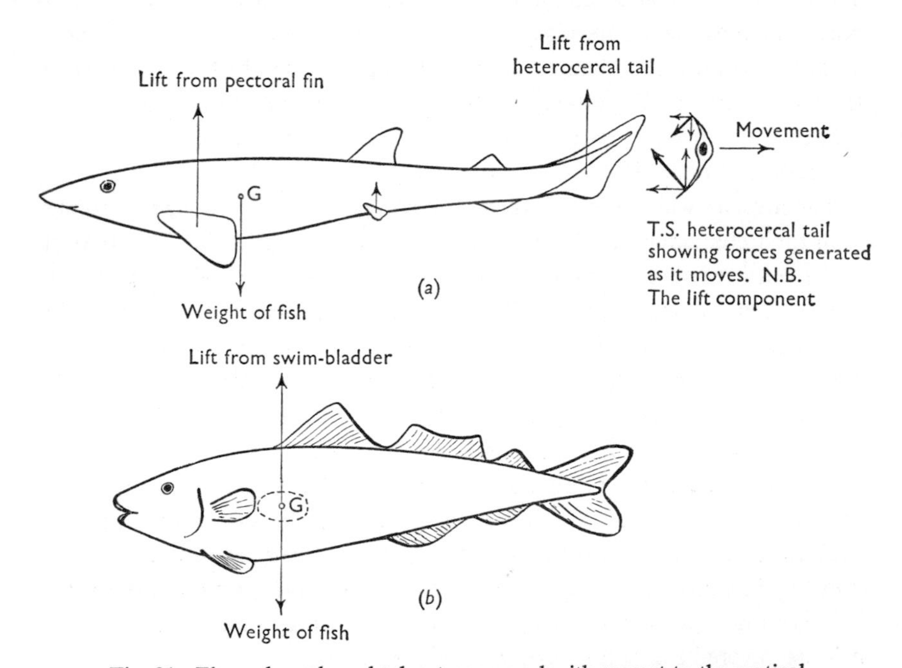

Fig. 91. Elasmobranch and teleost compared with respect to the vertical forces operating on their bodies. (*a*) Dogfish; (*b*) cod.

about its centre of gravity. These tendencies are corrected in fishes by fins which, though partly passive, also operate actively and so require the presence of sense organs (*semicircular canals of labyrinth*) which detect movements in these planes. In a dogfish all the fins, both paired and unpaired, tend to hinder rotation of the body about its longitudinal axis (i.e. roll). The tail and fins behind the centre of gravity tend to maintain the fish in its direction of motion but the unpaired fins in front of the centre of gravity counteract this tendency and make the animal more manœuvrable than if it were highly stabilised like an arrow. Although it is heavier than water a dogfish is able to swim on an even keel, partly because of the action of the pectoral fins and also because of the *heterocercal* tail. During its lateral movements this tail fin produces a lift force because the flexible caudal fin lags behind the stiff upper spinal column. The moment of this lift force about the centre of gravity tends to make the head tilt downwards but this is counteracted by an equal and opposite moment which results from lift developed at the pectoral fins (Fig. 91).

Bony fishes do not have heterocercal tails and most of them are not heavier than water. They maintain their position free-floating in water without apparent effort because of their *swim bladder*. This contains a bubble of gas whose volume can be regulated, thereby maintaining the density of the whole fish close to that of the sea- or freshwater.

7.62 *Terrestrial locomotion and its evolution*

In order to progress from aquatic to terrestrial locomotion, a number of changes were necessary in the skeletal and muscular systems. It has been supposed that during the conditions of the Devonian natural selection favoured these changes in a certain group of air-breathing fishes (the *Crossopterygii*) and from these the first terrestrial vertebrates—the amphibians—arose. This class of vertebrates, together with all subsequent groups, are known as *tetrapods*—the four-limbed animals—and their basic locomotory appendage is the pentadactyl limb (5-fingered). Although fossil evidence of the precise homologies of the bones is difficult to interpret, there can be no doubt that the pentadactyl limb arose from the pectoral and pelvic fins of Crossopterygii. Whereas in swimming the axial skeleton (i.e. the vertebral column) and segmental myotomes are functionally important, in walking and running the bones and musculature of the limbs and the limb girdles become of increasing importance both in support and propulsion.

When vertebrates first came on to land it was necessary to support their own body weight, and this effect of gravity profoundly influenced the whole of their subsequent evolution. Not only did the bones of the fin change to produce the pentadactyl limb, but major changes occurred in the vertebral column and limb girdles. The all-round support of the body provided by the aquatic medium was lost and replaced by a system of struts which transmitted the weight of the body through a chain of bones to the ground. For the support of the body it is clearly important that this chain of bones should be fixed to the main body axis. In fishes the pelvic girdle is embedded in the myotomes and has no direct connection with the vertebral column. In the early tetrapods the primitive plate of the pelvic girdle acquired a new dorsal process (the *ilium*) which became attached to the transverse process of the sacral vertebrae. In this way the main propulsive forces developed from the hind limb were transmitted directly to the axial skeleton of the animal. In the fore limb the girdle became detached from the hind end of the skull, which was its position in the bony fishes. This position was related to the region of weakness present behind the head because of the gills. Consequently, as this girdle became reduced and the gills were lost, a neck evolved enabling the head to have greater mobility, which was now adaptive as the need for smooth streamlining behind the head was no longer necessary. The pectoral girdle was embedded in the musculature and became shock-absorbing in function. This is highly developed in mammals, where the large scapula has serratus muscles on its inner face from which the body is suspended. Consequently, any shocks transmitted up the fore limbs are taken by the plate-like scapula and thence to the body through this muscular suspension. Their function is analogous to the mainspring of a car (see p. 147).

The vertebral column, which in fishes is markedly *metameric* (segmental repetition of parts) and facilitated lateral bending, now has quite different stresses acting upon the various sections along its length. In relation to this the uniform vertebrae of fish became differentiated into the five main types found in mammals (see above).

In early tetrapods, locomotion on land was accompanied by lateral undulations of the body. At this stage the limbs were unable to support the weight of the body, which consequently slithered along the ground. The movements of a newt or salamander are similar to those which might have occurred. In these amphibians the limbs are spread out *akimbo* and their movements are coordinated with the lateral undulations of the body. As soon as the animal stops walking it tends to collapse on its belly, yet despite its inefficient nature this means progressions persist in many reptiles

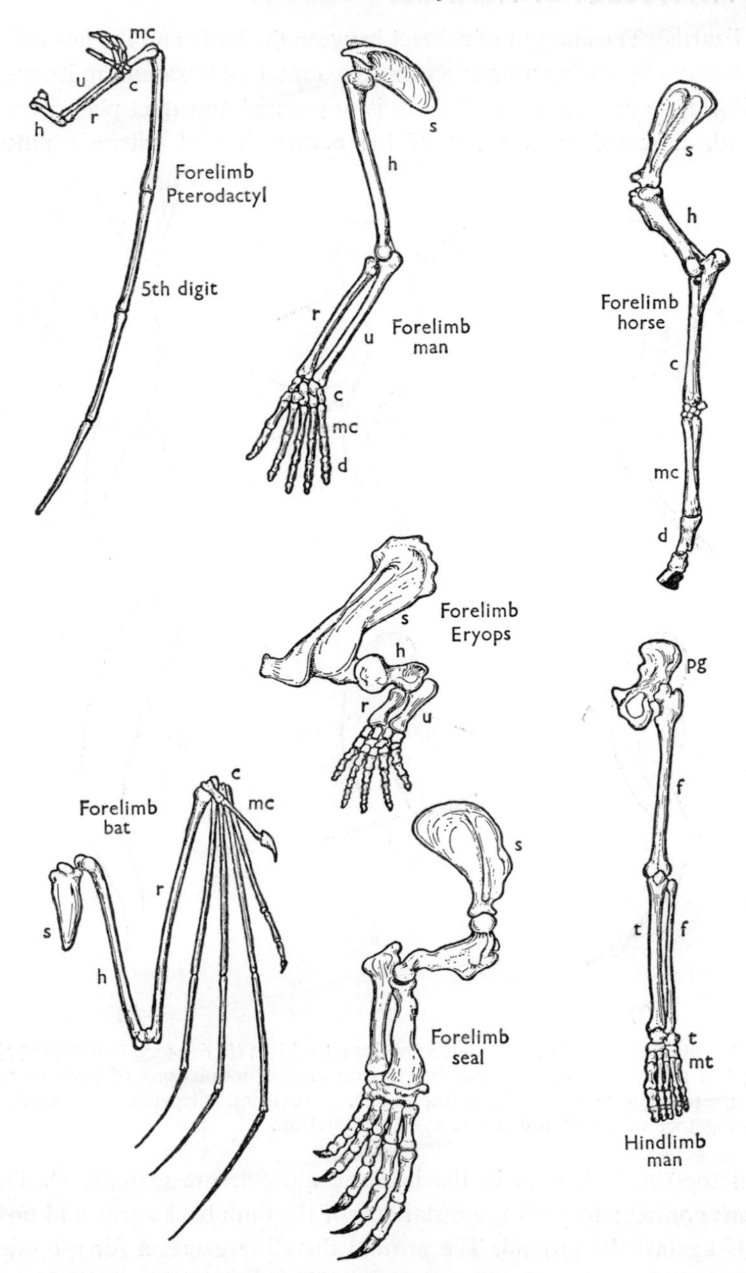

Fig. 92. Adaptation of pentadactyl limb for various forms of locomotion. s: scapula; h: humerus; r: radius; u: ulna; c: carpal; mc: metacarpal; t: tarsal; mt: metatarsal; d: digit; f: femur; t: tibia; fi: fibula.

(e.g. lizards). The amount of contact between the body and the ground can be observed by letting a newt move across mud and examining its tracks. During these movements each limb is protracted and then placed on the ground, retracted as a result of the contraction of extrinsic muscles

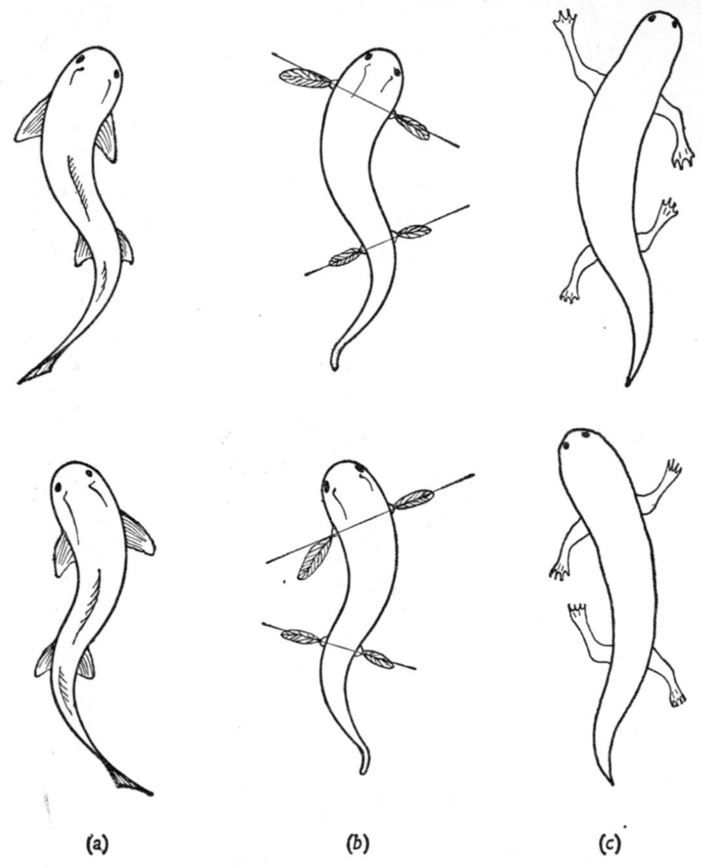

Fig. 93. The evolution of terrestrial locomotion. (*a*) Fish; (*b*) fish (e.g. Crossopterygian) using fins as passive extensions of body, with lateral undulations of back to assist movement on land; (*c*) primitive tetrapod (e.g. newt) using independent movements of limbs together with back muscles to assist locomotion.

(retractors) and changes in the intrinsic musculature (extensors). These actions combine to push the distal end of the limb backwards and downwards against the ground. The ground therefore exerts a force upwards and forwards against the limb and provided the hand or foot does not slip, this can cause the animal to move forwards (Fig. 93). The rhythm of the limbs with respect to one another in early tetrapods is again

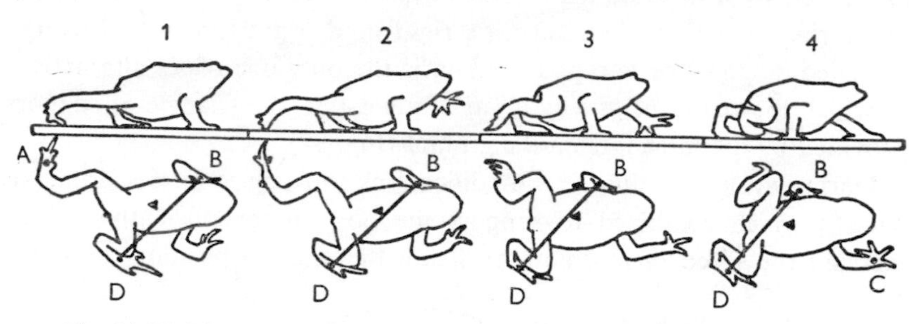

Fig. 94. Walking of toad. Notice how the centre of gravity moves over the stable tripod made by the feet which are on the ground.

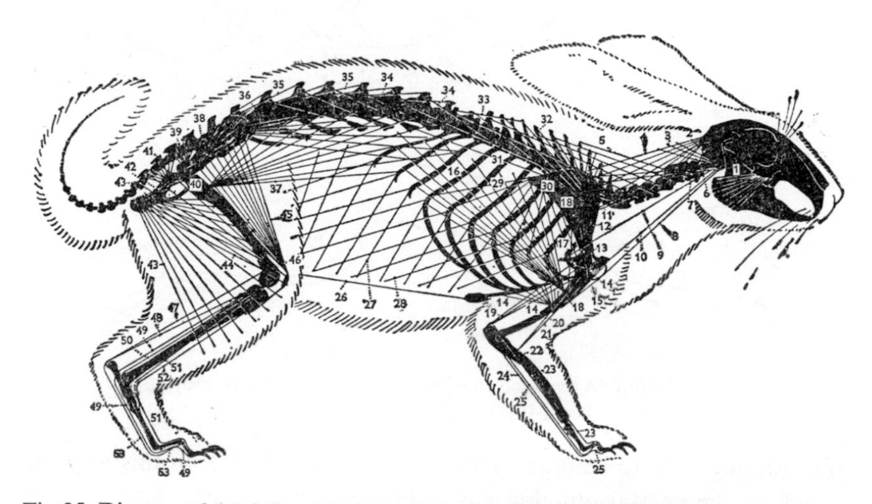

Fig. 95. Diagram of the skeleton and muscles of a rabbit, to show the general arrangement of struts and ties. 1, Masseter; 2, obliquus capitis; 3, splenius capitis; 4, semi-spinalis capitis; 5, longissimus cervicis; 6, longissimus capitis; 7, obliquus capitis inferior; 8, basioclavicularis; 9, levator scapulae; 10, sternomastoid; 11, scalenus; 12, supraspinatus; 13, infraspinatus; 14, pectoralis; 15, cleido humeralis; 16, latissimus dorsi; 17, subscapularis (displaced caudally); 18, deltoid; 19, triceps; 20, biceps brachii; 21, brachialis; 22, extensor carpi ulnaris; 23, extensor digitorum communis; 24, flexor digitarum sublimis; 25, flexor digitorum profundus; 26, rectus abdominis; 27, transversus abdominis; 28, external oblique; 29, serratus anterior; 30, trapezius; 31, iliocostalis; 32, longissimus; 33, semi-spinalis dorsi; 34, longissimus dorsi; 35, multifidus; 36, sacro-spinalis; 37, psoas major; 38, gluteus medius; 39, piriformis; 40, gluteus maximus; 41, abductor caudae; 42, gemellus inferior; 43, biceps; 44, adductors; 45, rectus femoris; 46, vastus intermedius; 47, gastrocnemius and plantaris; 48, soleus; 49, flexor digitorum longus; 50, peroneal muscles; 51, extensor digitorum; 52, tibialis anterior; 53, plantaris.

illustrated by a newt or toad when walking. The limbs are lifted in the order right fore, left hind, left fore, right hind, right fore, etc. This, the so-called diagonal or tetrapod rhythm, is the only one of six alternatives in which the centre of gravity always falls within the triangle of support provided by the three legs on the ground (Fig. 94).

During evolution numerous modifications took place in this primitive pattern which combined to support the body more efficiently and to increase the speed of locomotion. From the akimbo position the limbs

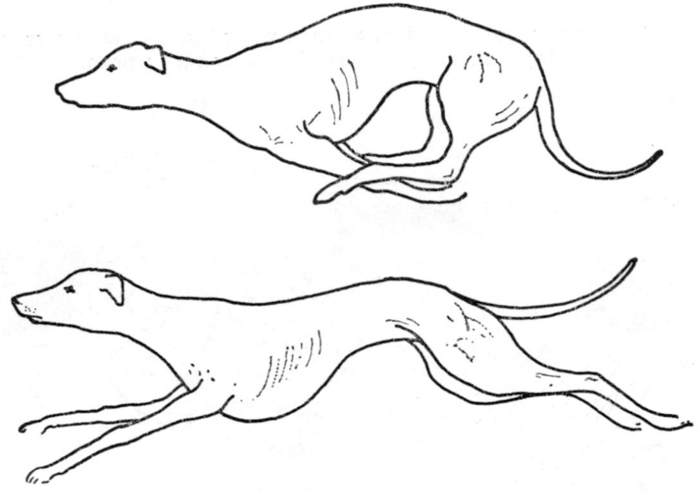

Fig. 96. Advanced tetrapod (e.g. mammal) using vertical movements of back, together with limb retraction.

rotated and came to lie more directly beneath the body. In this way the strain on the ventral muscles was reduced, and correlated with this the ventral elements (the coracoid) of the pectoral girdle became reduced. The weight of the body was transmitted through the chain of limb bones, which became lengthened and gave a longer stride, and enabled them to cover more distance per step. Such changes are present in modern mammals and are correlated with their mode of life. In many cases the increase in stride is due to the lengthening of the carpal and tarsal bones and lifting of the hind end of the hand and foot off the ground so that it walks on its digits (*digitigrade*). In other cases, such as the horse or cow, the number of digits is reduced and modifications at the toe produce a tough hoof, which minimises friction with the ground. The proximal elements of the limb are shortened and taken up into the body where they are surrounded by powerful extrinsic muscles. This has the effect of lightening the limb, which in the fleetest herbivores is little more than a chain of articulating bones and long tendons. Because of the more rapid movement and lengthening of the stride, changes in the gait occur. In some cases a greater length of time is spent on the diagonal part of the cycle (the trot) or on the phase with both ipsilateral limbs on the ground (the amble, e.g. camel) and at the fastest speeds there are phases when no legs are on the ground (the gallop).

In many of the fastest-running animals there are dorsal and ventral movements of the backbone which effectively increase the stride (Fig. 96). Notably this is found in the largest cats and deer; the cheetah is the fastest land mammal and at one phase of the cycle the hind limbs are in front of the fore limbs preparatory to their rapid retraction, which provides the main propulsive thrust. Aquatic mammals such as the whale reveal their mammalian descent by the up and down rather than side to side movements of their tails.

7.63 *Flying*

It is possible to distinguish two types of flight—passive or *gliding* flight and *active* or flapping flight. Many groups of vertebrates have become adapted for gliding but only birds and bats have evolved active flight. The essential structural requirement for flight is the possession of a membrane of *patagium*, which is an extension of the body which can support the body in the air. Birds are the only vertebrates in which the *patagium* is limited to the fore limb and has no connexion with the hind limb. This release of the hind limb has enabled it to become modified for a variety of functions, for example, swimming, climbing, running, etc. The patagial membranes of the bird fore limb are extended by the presence of feathers, which are a diagnostic character of these creatures. The whole body is clothed with them and besides their function in creating a smooth contour and in flight, they provide important insulation of the body which prevents the loss of heat.

7.631 *Feathers*. The feather is largely composed of keratin and develops as a light and strong secretion from an epidermal papilla. The basic structure is best described by considering one of the main *plume* feathers from the wing. Each of these is organised about a central stem or *rachis* which forms the quill of the feather upon which are inserted a number of *barbs* on each side. The barbs bear still smaller branches (the *barbules*) on both their proximal and distal faces. The distal barbules interlock with the proximal ones of the neighbouring barbs by means of microscopic *barbicels* so that an almost airtight membrane (the *vane*) is formed on either side of the rachis. Such plume feathers are found attached to the wings and to the tail. Other feathers which form the general surface of the body (*contour* feathers) do not have such complex interlocking of the barbs, for the small hooks are absent and the vanes are more fluffy. *Down feathers* are the

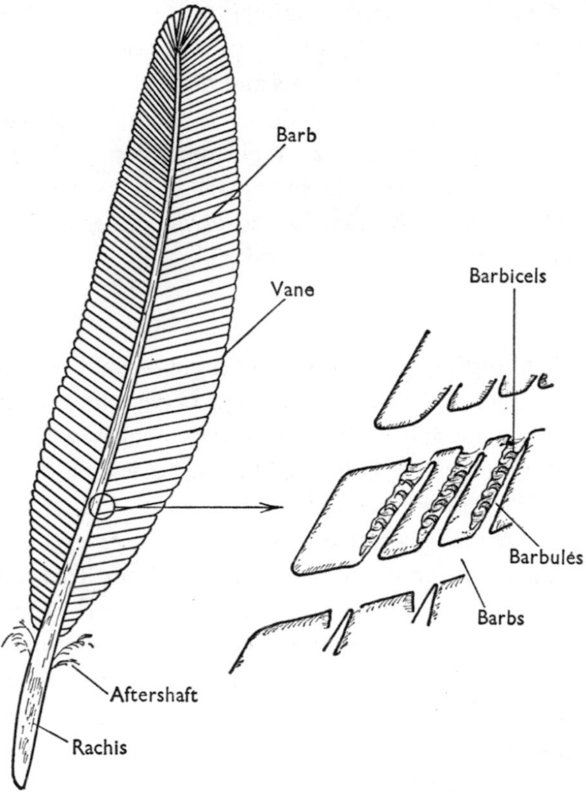

Fig. 97. Bird plume feather (see also Fig. 38).

simplest of the adult feathers and are loose and soft and may have a short rachis but are usually sessile. They are found between the contour feathers. The smallest types of feathers are known as *filoplumes*, which are very inconspicuous except during plucking. They are reduced feathers with a few barbs at the tip of a slender hair-like rachis.

Much of the bird's preening activity is directed towards sorting out these feathers and arranging them correctly.

7.632 *Structure of the bird's wing*. The way in which the basic pentadactyl limb is modified to form the wing of a bird is one of its most outstanding adaptations (Fig. 98). All the main segments are present—the humerus is large, the radius and ulna elongated, but the carpal and metacarpal regions have undergone fusion, reduction, and marked elongation. Only three digits are represented and these are now thought to represent the second, third and fourth rather than the first, second and third as is often stated. The anterior one (i.e. second) is short and bears the bastard wing. The

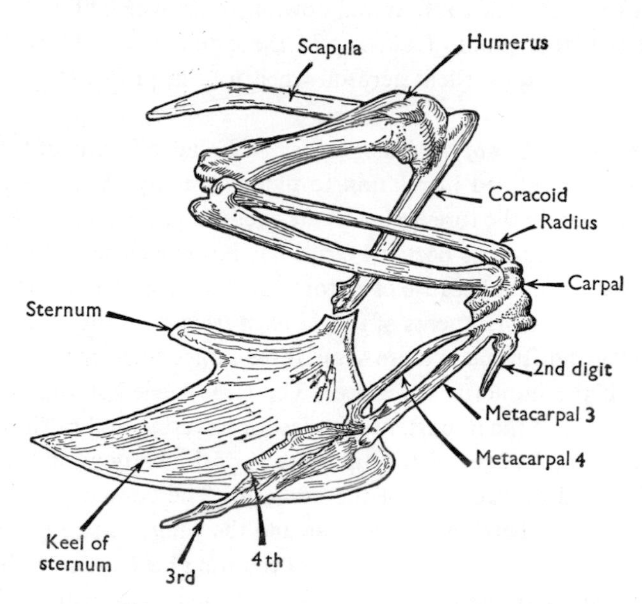

Fig. 98. Pectoral girdle and wing of bird.

middle digit (third) is the longest and has quite an elongated fused metacarpal region. From the functional point of view the wing is made up of the arm and the hand; the former bears the *secondary* feathers and the hand has the *primaries* attached to it. The structure of the plume feathers in these two groups may be differentiated by the relative size of the vane on the two sides of the rachis. These are equal in the secondary feathers but in the primaries they become more and more unequal as they spread round to the front or leading edge of the wing. The spacing of the feathers with respect to one another is maintained by a fenestrated membrane (*vinculum*) through which they are attached to the bone. Two important structural features of the skeleton are that when the elbow joint is extended the wrist joint automatically extends. Secondly, in the extended position the wrist cannot be rotated, which is very easily done in the flexed position.

Between the head of the humerus and the wrist and from the elbow to the hand are stretched the thin sheets of tissue which form the patagial membranes. That on the anterior edge is the most important and it is maintained in a stretched condition by the presence of special muscles (the tensor patagii muscles). The rest of the wing is very muscular and has in its skin a large number of contour and down feathers which fill in between the bases of the large plume feathers. In the extended condition the whole structure forms an excellent aerofoil when seen in profile (Fig. 99).

7.633 *Modifications of the limb girdles.* Although all parts of the skeleton of a bird are modified in relation to flying, notably by lightening of the bones, it is mainly the fore limb and the pectoral girdle which have become especially changed. The pectoral girdle has become closely attached to the large keeled sternum because of a stout coracoid bone. Dorsally this bone forms one of the constituents of the *foramen triosseum* (the other two being the scapula and furcula); these same three bones form the glenoid cavity with which the humerus articulates. The large keeled sternum forms the point of origin of the important muscles for lowering and raising the wing. The *pectoralis* (*P. major*) forms up to 25 % of the body weight in a pigeon and is inserted on the crest of the humerus. The coracoid forms a compression member between the sternum and the wing; its strength resists the forces exerted during the downstroke of the wing beat. Also with its origin on the sternum is the *supracoracoideus* (*P. minor*), which is very much smaller than the pectoralis (about one-fiftieth in a flapping flyer) and it sends a tendon through the foramen triosseum to be inserted on the dorsal side of the humerus. Thus two muscles with their origin and insertion on the same bones have opposite effects. Raising of the wing is normally accomplished by the pressure of the air and it is only during take-off and landing that the supracoracoideus muscle is involved to any extent.

The pelvic girdle of birds is greatly modified because of their bipedal habit. The ilium is fused very much with the sacral vertebrae which form the so-called *synsacrum*. Ventrally the ischia and pubis do not join with the members on the opposite side, so that there is no complete girdle in a bird. The tail is very much reduced in modern birds although it is elongated in early flyers such as *Archaeopteryx*. The hind limb of most birds is well provided with muscles which move it forwards and backwards and there are tendinous arrangements which ensure that the claws clasp a branch when the weight of the body is taken on the legs.

7.634 *Types of flight.* The principles of flight are the same for both gliding and flapping flight. When the wing is inclined at an angle up to 15° to the flow of air, it develops a lift force mainly because the more rapid flow

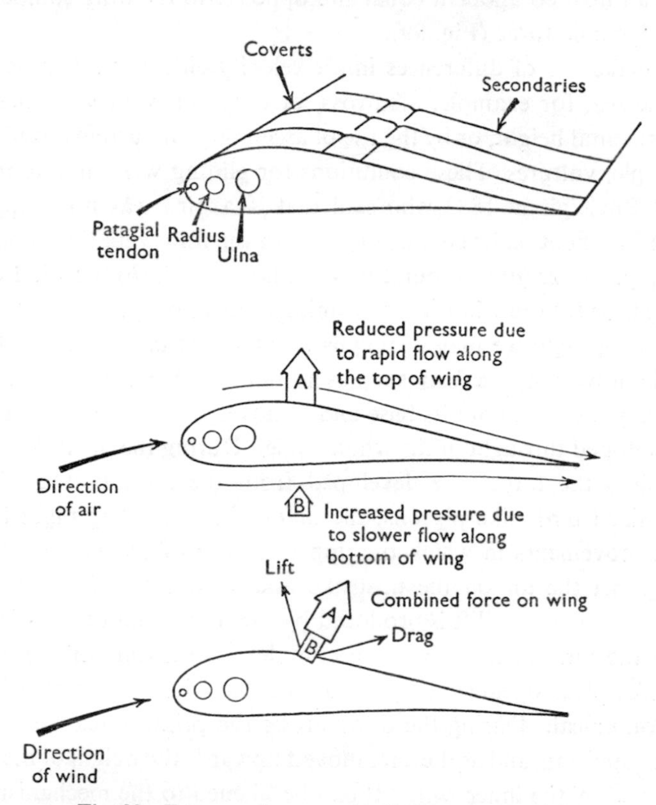

Fig. 99. Forces acting on a bird's wing during flight.

across its upper surface produces a region of reduced pressure (Fig. 99). The relative movement between the air and the wing may result from either movements of the air, as in rising air currents, or the loss of height of the bird, or by the active movements of the wing through the air. Under all these conditions there is a net resultant force on the wing which can be resolved into a component at right angles to the air flow (lift) and one in the same direction as the air flow (drag). When flying at uniform velocity the lift component must be equal and opposite to the weight of the body and the drag component must be opposed by another force driving the animal through the air. When a bird glides from the top of a tree to the ground in still air, the weight may be resolved into two components, one acting in a direction opposite to the lift component of the aerodynamic force, and the other acting in the direction of movement of the animal, which is a thrust component equal and opposite to the drag component of the aerodynamic force (Fig. 99).

Birds make use of differences in air velocity either at different heights above the sea, for example, albatross, or different wind velocities at the same horizontal height, or by the use of ascending air currents over deserts, for example, vultures. These conditions for gliding were first appreciated by Lord Rayleigh in 1883 who said that if a bird was not flapping its wings and yet kept on its course, then one of the following three conditions must apply. Either (*a*) the course was not horizontal, (*b*) the wind was not horizontal, or (*c*) the wind was not uniform in velocity.

In flapping flight we may distinguish that which takes place at take-off and in the hovering of a form such as a humming bird. In these birds the wing is very flexible about its base and it moves in a figure of eight so that lift is developed throughout the whole cycle. During the upstroke (relative to the bird) the forces are developed from the *back* of the wing. In medium-sized birds like pigeons, the take-off and landing flight involves complex movements in which the upper surface of the primary feathers strike against the air on the upstroke and so develop lift. During the downstroke, of course, lift is produced by the movement of the wing at an angle to the air. In fast flight the complex movements of the wrist no longer take place during the upstroke and the whole movement is much more economical. During the downstroke the primary feathers produce lift and propulsion, and as they are moved upwards the weight of the animal is supported by the inner wing. It can be likened to the mechanism of an aeroplane by saying that the primaries act as the propellers and the inner wing as the wings of the aeroplane.

In large birds it is not possible to make these complex take-off movements and they need to flap their wings when facing upwind. The noise of a swan taking off in the distance is well known.

NERVOUS COORDINATION

Each of the physiological processes described so far have their coordinating mechanisms which regulate their actions so that they are appropriate to the conditions prevailing at a given time. For this to be achieved systems of communication between different parts of the body are essential and vertebrates have two main means by which this is done: the first is by nervous impulses—electrical signals, and the second by hormones—chemical signals. These two types of signal are associated with the nervous and endocrine systems respectively.

In general the body needs to respond to stimuli from both the external and internal environments in order to produce adaptive changes leading to survival. Although the distinction is not exact, the nervous system tends to operate in the former case and the endocrine system in the latter. It should be realised, however, that the two are very closely linked both in origin and function and that the normal functioning of the body depends on the integrated action of the two.

For example, the nervous system depends for transmission between individual cells on the liberation of chemicals at their junctions with one another and also with effector organs, and nerve impulses can cause the release of hormones from endocrine organs (e.g. adrenals). In other words the two systems are not even distinct in the type of signal used. However, during the course of evolution the hormone or chemical signal has been selected to bring about slow and general changes in the metabolism of individual cells (such as growth), while the nervous system tends to be involved in rapid local responses to stimuli (such as movement).

The nervous system, with which we are concerned here, consists of specialised cells or *neurones*, linked together to form networks connecting organs which receive stimuli (*receptors*) and those which carry out actions or responses (*effectors*). It is an exceedingly elaborate system in mammals, with channels capable of rapid conduction (up to 120 m/sec for medullated nerve fibres) and a great specialisation and multiplicity of pathways between receptor and effector whereby a whole variety of responses is possible.

Before considering how the pathways of nervous communication are organised it will be as well to deal with the basic components of the system—that is, the neurone and its capacity to generate and conduct impulses, and the synapse across which the impulses must pass—as well as to discuss the origin of this physiological system from the embryonic tissues of the organism.

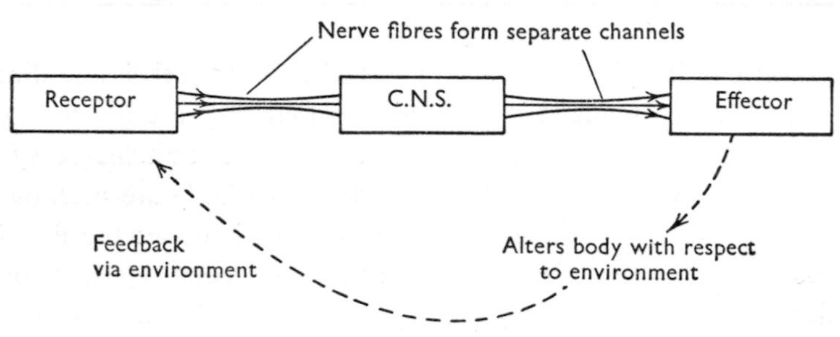

Fig. 100. The role of the C.N.S.

8.1 Structure and origins

It is characteristic of chordates (the phylum to which the vertebrates belong) that they have a hollow dorsal nervous system and this originates from the ectoderm which overlies the skeletal notochord. Other sources of nervous tissue are from the crests of the invaginating ectoderm (the *neural crests*), which give rise to collections of cells (the dorsal root and autonomic ganglia) outside the neural tube. In addition, local ectodermal thickenings associated with cranial nerves form *placodes* which sink in and give rise to other nervous structures which include the labyrinth and lateral line system as well as the olfactory organ.

The chief functional units of the nervous system are neurones, but there are other cells, called *neuroglia*, which make up as much as half the nervous system. They are generally thought of as supporting cells but they also play a vital role in the nutrition of the neurones. In all, the nervous system of man comprises some 3000×10^6 cells, a system which makes the most complex electronic computors appear simple in comparison.

8.2 Units of nervous function

8.21 *Neurone*

The structure of a typical neurone is shown in Fig. 102. Most neurones are characterised by a long protoplasmic process or *axon* which is specialised for the conduction of impulses from one part of the cell to another. Neurones which conduct impulses from receptor organs to the central nervous system are called *sensory* or *afferent*, while those conducting impulses outwards from the central nervous system to effectors are *motor* or *efferent* neurones. The former have their cell bodies outside the spinal cord in a ganglion of the dorsal root while the motor neurones have their cell bodies within the spinal cord and emerge via ventral roots.

It is possible to subdivide the sensory and motor systems into *somatic* and *visceral*, the first being concerned with coordination of the skeletal or striated muscle systems and the second with the control of visceral activities such as the smooth muscle of the gut. The visceral system, both motor and sensory,[1] mainly coordinates functions of the body below the level of consciousness and this forms the *autonomic system.* It will be discussed more fully later in this chapter.

Despite the large number of afferent and efferent neurones, the vast bulk of the C.N.S. is made up of *interneurones* which lie entirely within the brain and spinal cord. These tend to have branched axons: some may be very long and form extensive tracts which pass up and down the spinal cord, but others have very short axons and only communicate over short distances.

In structure the neurone has a cell body or *soma* which is the main nutritional part of the cell and is concerned with the biosynthesis of materials necessary for the growth and maintenance of the neurone. The cell body has a large nucleus and many mitochondria and other granules (Nissl granules) associated with synthesis and energy exchange. In some cases, for example, dorsal root ganglion cells of mammals, the cell body plays little part in the transmission of impulses as it is on a side branch of the neurone. In most cases, such as motor neurones, the cell soma is the site of many endings from other cells which form 'end buttons' all over its surface. Whether the cell will be excited or not is determined by the summated effect of all these endings, some of which may tend to excite but others to inhibit the production of a nervous impulse. In motor neurones the nervous impulse arises at the point (the axon hillock) where the axon takes its origin from the soma. Other regions of neurones form very fine branching processes known as *dendrites*, and these are especially numerous in interneurones. These are again the points where different nerve cells come into contact with one another and communicate information between their surfaces.

[1] See footnote to p. 211.

8.22 *Nerve impulse*

By the use of minute electrodes (e.g. glass capillary micro-electrodes of less than 1 μ tip diameter) inserted into the soma or axon a great deal has been discovered about the nature of the nerve impulse. It has been found that most excitable tissues, whether nerve or muscle, have a difference of electrical potential between the inside and outside of their external membranes. This *resting* or *membrane potential* is such that the outside is positively charged with respect to the inside.

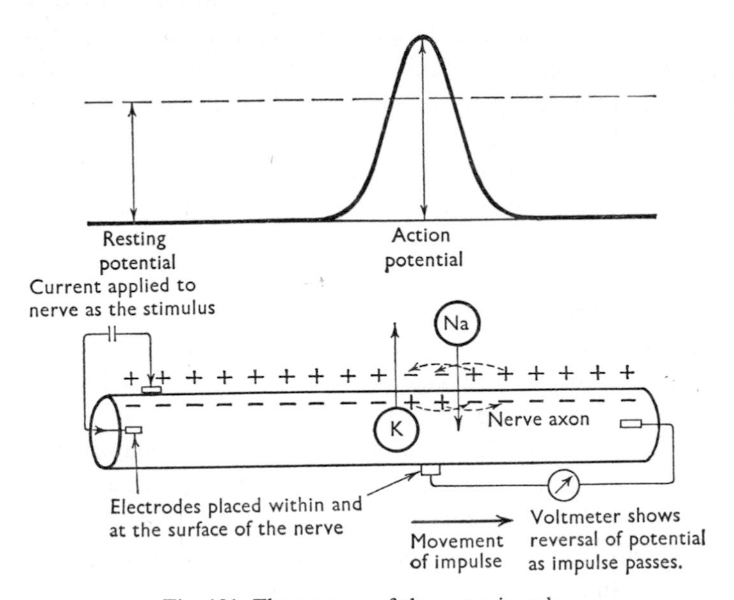

Fig. 101. The passage of the nerve impulse.

These facts were first established using giant fibres of the squid, which may be up to 1 mm in diameter, by placing electrodes on each side of the membrane and connecting them to a sensitive galvanometer. The readings showed that there was a difference in potential of 70–90 mV (0·07–0·09 V) across the membrane and it is largely due to differences in the concentration of K^+ ions inside and outside the cell. (The concept of the Na^+ pump is dealt with on p. 124, under 'ionic regulation'.)

On stimulation of the neurone, a change takes place in the potential recorded by the galvanometer whereby the charge on the outside is reduced and finally reversed. As the outside now becomes negative relative to the inside, current flows from the resting part of the nerve into this region and consequently sets up small local circuits which themselves excite the neighbouring regions of the axon and hence the impulse passes away from the initially excited region (Fig. 101).

This changing potential which passes along the nerve fibre and thus corresponds to the passage of an electric impulse has been shown to be due to an inrush of Na^+ ions across the membrane followed by an outrush of K^+ ions. The sodium pump mechanism removes the Na^+ ions that have entered and soon repolarises the surface of the membrane as in the resting state.

Although a neurone can be caused to generate an impulse anywhere along its length by an appropriate stimulus (direct stimulation of the nerve fibre passing across the 'funny' bone of the elbow by a blow causes pain) the impulse normally originates at the cell body or the sensory endings. The propagated impulse is preceded by a *generator potential*, and if this does not reach sufficient amplitude impulses do not arise; that is, weak stimulation does not lead to smaller impulses. This phenomenon is associated with the all-or-none character of the response of the neurone to stimulation. Increase in the stimulation at a sensory ending when it is above threshold leads to an increase in the frequency of the impulses that are generated and not a change in their individual nature.

Following the passage of a nervous impulse the axon membrane is inexcitable for a time known as the *refractory period* (about 1 millisec). Thus there is a limit to the frequency of impulses that nerve fibres can transmit which is usually 500–1000 impulses/sec. To summarise, therefore, the nerve impulse seems to be a wave of negativity generated in special parts of the cell and passing down the whole length of the axon; it is always the same size from a given neurone and varies only in the frequency at which it is propagated. The rate of transmission depends on the cross-sectional area of the fibre and the nature of its external covering. In vertebrates the conduction velocity of the impulses is increased because of a specialisation of the outer sheaths of the axon called the *myelin sheath*. Nerve fibres which have this sheath, called *medullated*, conduct more rapidly than those without (the *non-medullated*). Figure 102 shows the structure of the medullated fibre and it can be seen that the sheath is interrupted at various places called the *nodes of Ranvier* where the axonal membrane is exposed to the tissue fluid. The sheath cells are rich in lipoid and effectively insulate the fibre so that it is only excitable at the nodes.

It has been shown in fact that the impulse is conducted from node to node. This saltatory or jumping conduction enables the velocity of conduction to be much increased without enlarging the diameter.

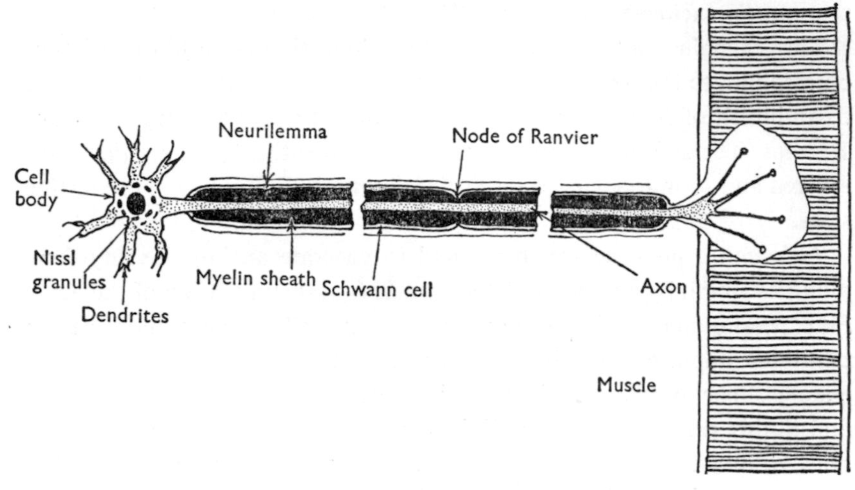

Fig. 102. Motor neurone.

8.23 *The synapse*

We must now consider what happens when the nerve impulse reaches the end of the axon along which it has been conducted. The endings may be associated either with an effector organ such as a muscle fibre or gland cell, or if the axon belongs to a sensory or interneurone it will make connexions with other neurones. Regions where two neurones are functionally connected are known as *synapses*. They represent discontinuities of both structure and function. In many ways the endings of motor axons on muscle fibres are similar and have been shown to have properties in common with those of true nerve-to-nerve synapses. In all of these cases recent electron microscope studies have shown a very close approximation of the two membranes of the excitable cells (100 Ångström).

Despite this very short distance for transmission of the impulse, there is undoubtedly a relatively long *delay* (0·5–0·9 millisec) between the arrival of the nerve impulse in one fibre (*pre-synaptic fibre*) and the setting up of an impulse in the *post-synaptic fibre*. Part of this time is taken up by processes whereby small amounts of specific chemical transmitters are liberated which then depolarise the post-synaptic membrane which leads to the setting up of an impulse in that neurone. *Acetyl choline* is the best known of these transmitters and there is no doubt that it operates at the neuromuscular junction of vertebrate skeletal muscle. One simple experiment showing this is the effect of poisons which render the post-synaptic membrane insensitive to acetyl choline and hence block transmission. Such a poison is *curare*, which has long been known as a poison used by South American Indians on their arrowheads. After the acetyl choline has been liberated and has produced its effect at specific sites on the post-synaptic membrane, it is destroyed by enzymes called *cholinesterases*. Many drugs are known which specifically prevent the activity of these enzymes and

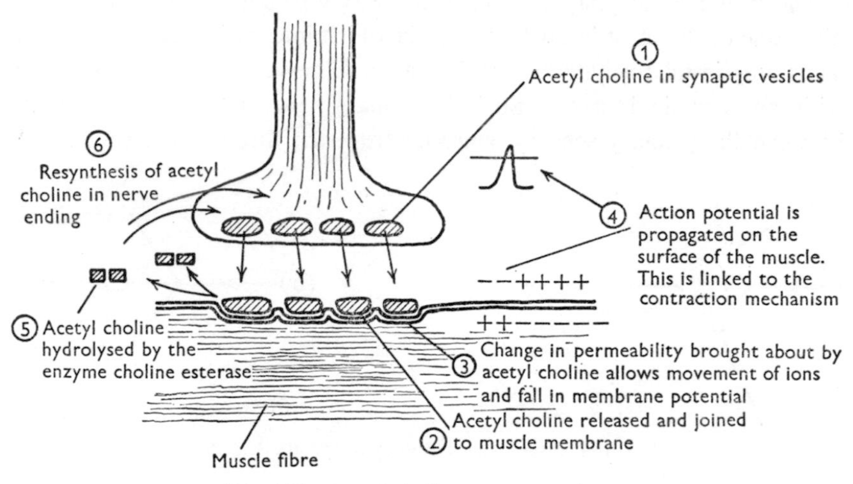
1
Acetyl choline in synaptic vesicles
6
Resynthesis of acetyl choline in nerve ending
4
Action potential is propagated on the surface of the muscle. This is linked to the contraction mechanism
– – + + + +
+ + – – – – –
5
Acetyl choline hydrolysed by the enzyme choline esterase
3
Change in permeability brought about by acetyl choline allows movement of ions and fall in membrane potential
Acetyl choline released and joined
2
to muscle membrane
Muscle fibre

Fig. 103. Acetyl choline as a transmitter

consequently have a marked effect on transmission at the neuromuscular junction. Well known amongst these anticholinesterases are eserine and some of the poison war gases such as D.F.P. (diisopropyl fluorophosphonate).

Transmission at the neuromuscular junction has been investigated in great detail because of its ready accessibility in physiological experiments. But similar studies, such as the use of drugs, have been applied to many parts of the C.N.S. and to ganglia of the autonomic system. For instance, if the presynaptic fibres of a sympathetic ganglion are stimulated, while it is being perfused, acetyl choline can be detected in the fluid leaving the ganglion during stimulation. This is most clear if the perfusing fluid also contains an anticholinesterase such as eserine. Notable among the properties of synapses is their *polarity*. That is, the impulse can only cross the synapse in one direction. Hence, although it is possible under suitable conditions for a nerve impulse to pass both ways along an axon, when it reaches a synapse it only produces an effect on another neurone or motor cell in one direction. These differences are due to the properties of the two membranes on both sides of the synapse. Polarity within the nervous system of vertebrates contrasts with that of the nerve nets of primitive invertebrates in which impulses apparently can pass both ways across their synapses.

8.24 *Integrative mechanisms*

The nature of the synapses and the detailed connexions between neurones gives one of the most important ways in which integration can take place. Thus only in relatively few cases does a single nerve axon make a synapse with another single axon. Such 1:1 synapses are found, for example, between the primary sensory neurones from the *fovea* of the retina and

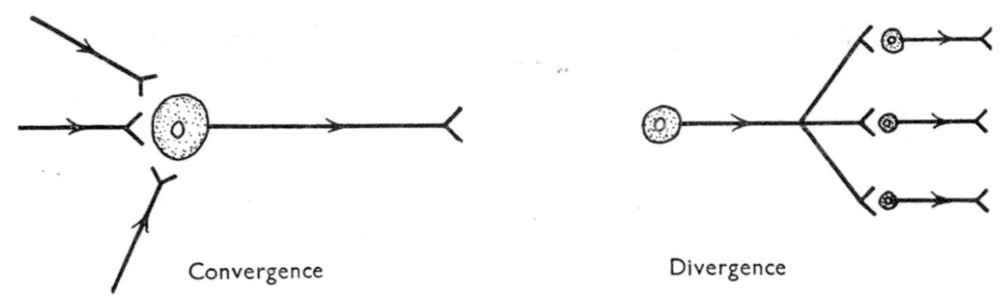

Fig. 104. Two sorts of connexions in the C.N.S.

first-order interneurones. In such cases the synapses are simply relay stations, each impulse in the pre-synaptic fibre producing one in the post-synaptic element and there is no integration. More usually, however, several pre-synaptic fibres have their endings on a single postsynaptic neurone and in these instances the arrival of a single impulse in one axon is insufficient to excite the post-synaptic cell, presumably because it does not liberate a sufficient quantity of the transmitter substance. If another impulse arrives very soon after in either the same axon or, more usually, along other axons ending on the cell, then the post-synaptic element may be excited. This type of connexion is illustrated in Fig. 104 and is called *convergence.* The effects of impulses arriving along different pathways or the same pathway are said to *summate* with one another. It is also common to find in the C.N.S. that impulses passing along a single axon may produce effects on a large number of post-synaptic cells. This is called *divergence,* and well-known examples are found among the giant fibre systems of invertebrates. Among vertebrates, analogous cells are found in fishes and some larval amphibians in the spinal cord where they are effective in initiating sudden swimming by the animal. But such divergent connections have been established in many functional tracts descending from the brain of mammals to the spinal cord.

In considering the C.N.S. in this way it is inevitable that a picture arises of nerve cells only responding to the arrival of impulses in preceding members of the chain, and may be presumed to have originated ultimately from the receptors. That this is not necessarily true is now well established because many nerve cells throughout the animal kingdom are known to have activity in the absence of any preceding input in the way of nervous impulses. We have seen a clear example of this 'spontaneity' when considering the respiratory centre of mammals and fishes, for even when the medulla is completely isolated some cells continue to discharge with frequencies and patterns almost identical to those which accompany the normal respiratory movements of the intact animal. Again the fore brain of vertebrates shows clearly defined electrical activity which in mammals is the well-known alpha rhythm (10/sec).

8.3 Receptor organs

8.31 *Basic mechanisms*

In experimental work with the nervous system it is usual to employ small electric shocks for stimulating the axons. This is because such stimuli are easily measured and varied in frequency and intensity. But nerve fibres and cell bodies can also be excited by mechanical, chemical, light, and temperature stimuli. This applies particularly to some of the branching endings found in the skin of primitive organisms. Such receptors play a role in detecting changes in the environment of the animal but the information they communicate to the C.N.S. is not very specific because it may have arisen from any of these types of stimulation. During evolution there has been a division of labour among receptor organs so that they became more and more specialised in the type of stimulus or *modality* to which they respond. This is usually due to changes in *accessory structures* associated with the endings of the receptor neurones. The function of this apparatus is to protect the endings from certain types of environmental change and to concentrate others. In this way a high degree of specificity and sensitivity of the receptor mechanisms has evolved which can give extremely detailed information about both the internal and external environments.

Once a particular receptor neurone is excited, impulses are transmitted along its axon to the C.N.S. All impulses originating in a given receptor cell are identical but they vary in the frequency with which they are generated. Frequency is therefore the code and the relationship between stimulus and frequency is a characteristic property of each receptor. When a sense organ is suddenly stimulated, for example, by shining a light or suddenly exerting a pull on a muscle, the initial frequency of sensory

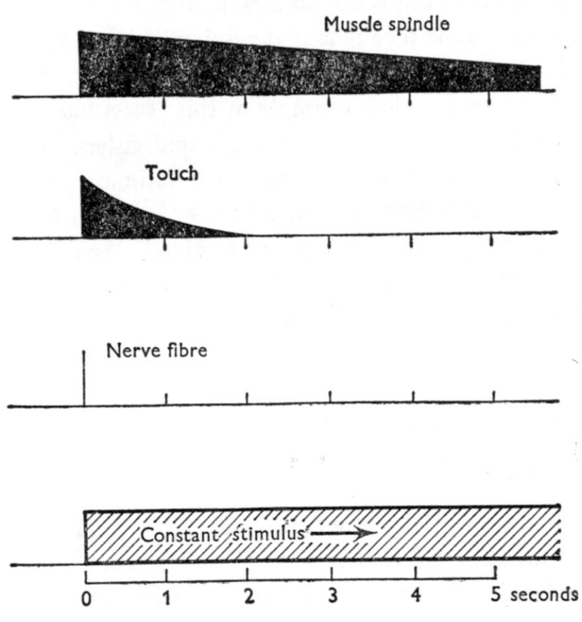

Fig. 105. Change of frequency of nerve impulses with time to a constant stimulus. The nerve fibre is medullated.

impulses is very high but falls away with time, although the stimulus intensity remains constant. This property is fundamental to all sense organs and is called *adaptation.* In some cases the frequency falls very rapidly (Fig. 105) and these are said to be *rapidly-adapting* sense organs. Examples are those concerned with touch and the detection of sudden movements. Other receptors are characterised by a relatively slow decline in the discharge frequency—*slowly-adapting*. Such receptors play a vital role in indicating the more or less steady conditions of the organism with respect to its internal and external environments. Examples are the receptors in muscles which detect their tension, or those in a joint which signal the angle at which it is held. In all sense organs the intensity of the stimulus, for example, the brightness of a light or the amount of pull exerted on a muscle, determines both the maximum initial discharge frequency in the sensory fibres as well as its level at all stages in the adaptation curve. In sense organs where there are more than one receptor cell, which is not uncommon, different cells may become excited at different intensities; hence the C.N.S. receives a variable inflow not only in frequency of impulses along a given axon but also in the number of axons transmitting impulses. Each sensory neurone has connections within the C.N.S. which, because of their nature, give rise to sensations of a particular type. In whatever way the fibres are excited, however, the same sensation is experienced subjectively. A well-known example is the way pressure on the eye-ball produces the sensation of light just as if the optic nerve fibres had been stimulated by a light stimulus.

8.32 *Classification*

Numerous ways are possible for classifying the different types of sense organ; a common one depends on the site of the stimulus which excites the organ. On this basis we may distinguish *exteroceptors*, which detect changes in the external world, and they may be divided into:

teloreceptors or distance receptors,
cutaneous receptors, and sometimes
chemical receptors, which include the senses of taste and smell.

Interoceptors are stimulated by changes in the internal environment and are divisible into:

proprioceptors, stimulated by the position or activity of the body itself,
visceroceptors, stimulated by part of the viscera, and a third group,
chemical interoceptors, such as those detecting the O_2 tension in the carotid glomi.

In another system of classification the nomenclature is based upon the type of environmental change or energy to which the receptor is particularly responsive. On this basis we may distinguish:

chemoreceptors (smell and taste),
mechanoreceptors (touch, pressure, hearing, etc.),
photoreceptors (light, e.g. eyes),
thermoreceptors (temperature) and
undifferentiated endings, which produce the sensation of pain.

In certain fishes it is apparent that there are also receptors which are sensitive to the deformation of small *electric* fields produced by the fish itself.

Whatever the type and classification of a receptor the basic mechanism is the same. The environmental change is concentrated in some specific way upon nerve endings which produce small potential changes which in turn generate nervous impulses that are transmitted to the C.N.S. In all cases the essential change which precedes the *generator potential* is either chemical or mechanical. Thus, in the eye there is a lens and devices to focus an image on the receptive retinal cells where photochemical changes take place. The chemical events excite the receptor cells to produce a slow change of electrical potential which in turn generates nervous impulses.

8.33 *Mechanoreceptors*

8.331 *Touch.* Recordings made from small nerves supplying mammalian skin show that both slowly and rapidly adapting receptors are present. The former are suitable for the detection of touch and vibration, whereas the latter give information about continuous pressure on the skin. There are at least three different types of sensory ending involved. The first are associated with the base of hairs, particularly on the 'windward side'. These receptors are extremely sensitive to small movements, especially in the vibrissae of some mammals. Some can detect movements produced by air currents and in this way act as teloreceptors. A single nerve fibre may serve endings at the base of hairs over an area as great as 5 cm^2. A second type of ending is found on relatively non-hairy regions such as the fingers where the main cutaneous sense is due to Meissner's corpuscles which lie in papillae which extend into the ridges. The corpuscles consist of spiral and much-twisted endings each of which ends in a knob. A third type of end-organ is the very sensitive and large Pacinian corpuscles which are usually quite deeply situated and in the limbs may lie close to the tibia; they are also well known from the gut mesenteries. They have been very well investigated physiologically and those in the limbs probably form the basis of a vibration sense, for they are very rapidly adapting as well as being highly sensitive; for example, a deformation of $0{\cdot}5\mu$ if applied for 100 millisec will excite them. Their rapid adaptation has been shown to be due to the properties of the complex accessory corpuscle arrangement which protects the end organs.

The precise relationship between end organs observed in sections of skin and the different sorts of response recorded from cutaneous nerve fibres has not been established in all cases. As indicated above a single sensory fibre may be stimulated by movements of hairs over a relatively large area and the responses are rapidly adapting, but touch receptors are generally more slowly adapting and their receptive areas may be limited to one or more very discrete spots on the skin. Pressure receptors have a much higher threshold and have no sharply localised receptive spots. That pressure reception is due to a deformation of the receptor by pressure gradient of the skin, and not pressure as such, is indicated by a classical experiment of Meissner. If a finger is thrust into a vessel of mercury the sensation of pressure is only obtained at the interface where the skin is deformed by the pressure gradient, there being no sensation of pressure within the mercury.

A great deal of work has been done on the so-called 'two-point threshold', that is, the determination of the minimum distance at which two mechanical stimuli can be distinguished from one another. This distinction can be very fine (one or a few millimetres) in certain parts of the body such as the tips of the fingers, the tongue, and the lips, but in other parts it is relatively poor; for example, on the thigh, arms and neck it may not be possible to distinguish single points applied to the skin at distances of 5 cm or more apart. At first sight it might be thought that the basis for this discrimination would depend upon the size of the areas of skin supplied by the sensory endings. This is certainly involved, but the mechanism is more complex because the areas which may be discriminated are smaller than the minimum areas innervated by a given sensory fibre. The fields of these sensory endings overlap one another and the pattern emerging from their stimulation is one factor involved in the integration mechanism.

8.332 *Proprioceptors.* These often somewhat neglected receptors are generally slowly adapting and are represented in the muscles of amphibians and mammals by the *muscle spindles* and tendon organs. Each spindle consists of a few muscle fibres (*intra-fusal*) with contractile striated portions at their two ends and between which are the sensory endings. The spindles are therefore in parallel with the main mass of the contracting muscle fibres. On the other hand, *tendon organs* are in series with these fibres. When a muscle contracts and its tension increases, the tendon organs will be excited but only at a relatively high intensity or threshold and they do not play such an important part in the regulation of body movements as do the muscle spindles. The tendon organs are stimulated during the 'clasp-knife' reflex which results in the inhibition of an extensor muscle when it is stretched beyond a certain amount. It protects the muscle from damage due to overloading. The afferent fibres from the muscle spindles are large (20 μ) and have a series of spiral turns round the intra-fusal fibre. By recording from their nerves it has been shown that these endings are excited either when the muscle is stretched passively, or if the intra-fusal fibres themselves contract. The discharge is slowly adapting and increases in frequency the

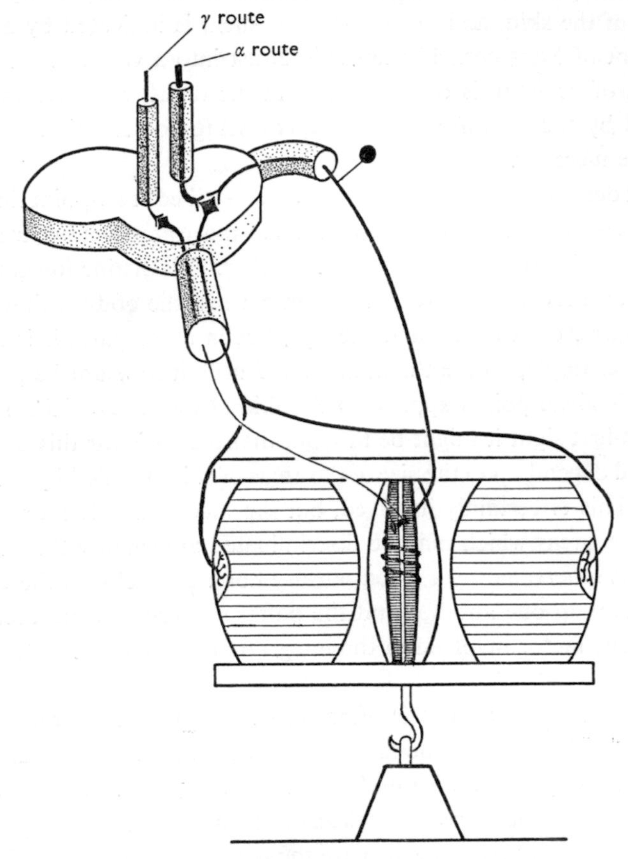

Fig. 106. The use of proprioceptors (muscle spindles) to produce an appropriate degree of muscle contraction to meet a given load. γ route is a slow path; descending fibres from the brain excite motor neurones whose fibres pass out of the ventral root to stimulate the muscle spindle. The latter contracts and sends a stimulus via the spinal arc to the muscle. This stimulus, and therefore the contraction, will be appropriate to the load. α route is a fast 'emergency' path from the brain via the spinal cord directly to the muscle motor nerves. It results in a rapid, but non-graded, contraction.

greater the stretch of the muscles. As will be discussed later, these endings form a vital part of the stretch reflex (see p. 207), which plays an important part in regulating the posture and movements of mammals.

If the motor nerve fibres to the ordinary muscle fibres are excited electrically, the muscle shortens and the discharge of impulses from a muscle spindle is interrupted, and in this way reflex excitation of the motor neurones is reduced during the contraction. The contractile regions of intra-fusal fibres are also innervated and have end-plates at both ends. These motor nerve fibres are relatively thin (gamma fibres, 3–8μ in diameter) and when excited they increase the tension within the muscle spindle and hence augment its discharge frequency. This motor system is therefore able to bias the receptors in definite directions. One way in which it plays an important role (Fig. 106) is during slow postural movements when the γ efferents are excited by descending pathways from the brain. Contraction of the intra-fusal muscle fibres will then alter the input from the spindles in relation to the load that the muscle is working against. Consequently, the ensuing reflex excitation of the main motor neurones of the muscle will be greater or less according to this load and the resulting contraction will be graded to overcome the load itself. This example illustrates the importance of proprioceptors as parts of *feedback loops* in the control of muscular activities. The fact that the muscle spindle afferent fibres are large in diameter and rapidly conducting ensures that delays in this control loop will be brief. It is useful in reflex experiments because electrical stimulation of the sensory nerves excites these large fibres at the lowest threshold. Because of this the mechanisms involving the small motor fibres have only been elucidated relatively recently.

8.333 *Other interoceptors.* Within the vertebrate body there are many receptors which respond to the mechanical conditions of the internal organs. The input from these receptors often plays an important part not only in the control of particular physiological mechanisms, but it may also influence certain behaviour patterns of the whole animal. For example, the receptors of the stomach wall may be concerned in the arousal of 'hunger'. Stretch receptors in the carotid and aortic sinuses of tetrapods have important roles in the regulation of blood pressure. Endings with similar properties are found in the branchial vessels of fishes.

8.334 *Vibration sense and lateral line organs.* These receptors are intermediate between mechanoreceptors which detect movements of the body and those described as hearing organs which detect higher frequencies of air- or water-borne vibrations. The lateral lines of fishes and some amphibians are formed of canals just beneath the surface of the skin which open periodically to the exterior. At intervals within each lateral line canal, there are groups of sensory hairs called *neuromast organs*. These sensory hairs are innervated by fibres which run mainly in the Xth, VIIth and Vth cranial nerves and enter the medulla. Information from them is therefore gathered together in one region of the C.N.S. although the receptors are distributed widely over the whole body surface. Recordings from lateral line nerves showed the

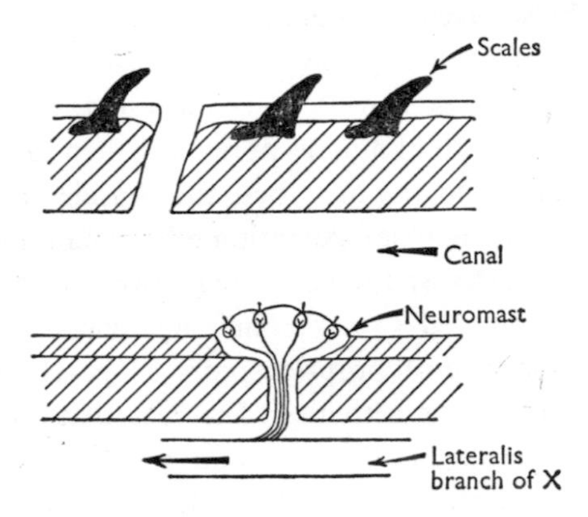

Fig. 107. Longitudinal section of a lateral line of dogfish.

presence of a resting discharge, accelerated by the passage of water along the canal in one direction and decreased by flow in the reverse direction. In the ray an increase occurs when perfusion is from head to tail. The lateral line system is sensitive to vibrations in water of up to about 100/sec and is concerned with the detection of objects moving near to the fish and also of objects into which the fish might swim. As a fish approaches some obstacle or object the lateral line system will detect any consequent alterations in the pattern of water flow over its surfaces.

The detections of vibrations of the substratum by terrestrial vertebrates is probably achieved by receptors at the joints. It is clear then that the division between movement receptors, proprioceptors and other mechanoreceptors involved in hearing, etc., is not a sharp one for they all grade into one another. Because of its separation as an organ of special sense, however, it is convenient to discuss the ear separately, although it forms part of this continuous spectrum.

8.34 *Sound and equilibrium receptors*

8.341 *The ear—equilibrium reception and hearing.* The inner ear of vertebrates develops from a *placode* (epidermal thickening) to form a hollow otocyst which subsequently differentiates into two parts, the lower (*pars inferior*) forming the *sacculus* and *lagena* (*cochlea* in higher tetrapods) and an upper region (*pars superior*) composed of the *utriculus* and three semicircular canals (only two in some cyclostomes). The whole structure represents a portion of the lateralis system which has become closed off from the exterior except for the *ductus endolymphaticus*, which remains

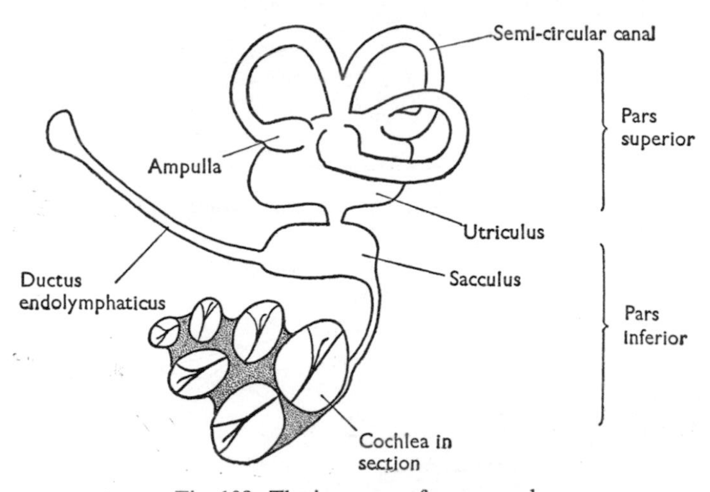

Fig. 108. The inner ear of a mammal.

open in fishes. The semicircular canals have small groups of hairs corresponding to the neuromast organs which detect movement of the fluid (*endolymph*) within the labyrinth. Each canal is oriented at right angles to the others and its neuromasts respond to angular accelerations in the plane of the canal. The neuromast organs are contained within the ampullae at one end of each of the canals. Their sensory hairs are embedded in a mucilaginous *cupula* which functions as a pendulum that is swayed by movements of the endolymph (Fig. 109).

Neuromasts are also modified to form *maculae*. Their sensory hairs are embedded in mucilage which contains calcareous bodies or *otoliths* which may be quite large in some fishes. The degree of stimulation of the sensory hairs of the maculae depends on the position of the labyrinth and they therefore provide tonic information essential for orientation of the body with respect to gravity. They will also respond to linear acceleration of the body because of the inertia of the otoliths (concretions of substances such as $CaCO_3$) used in gravity responses. The difference in density of the calcareous bodies and surrounding medium also means that the otolith will vibrate when subjected to sounds. In fishes the maculae are known to detect vibrations of higher frequency than those which

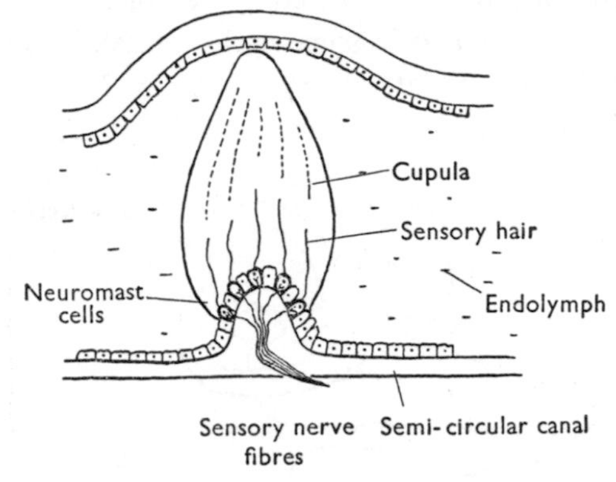

Fig. 109. Section of an ampulla.

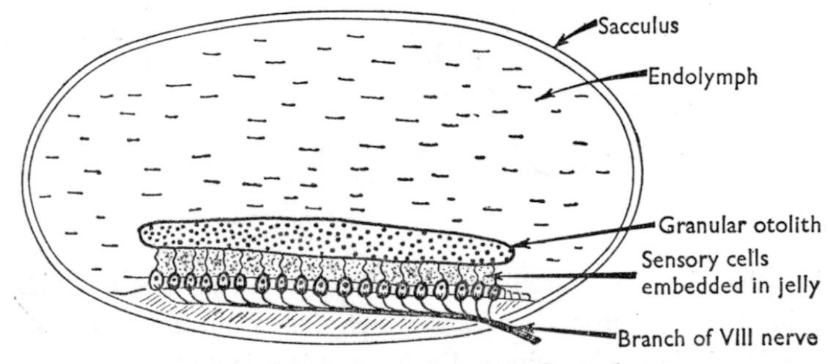

Fig. 110. Transverse section of macula.

stimulate the lateral line organs. Some fish can detect even higher frequencies (10,000/sec) by the use of the swim-bladder wall, which vibrates when sound waves are passed through the body. In fish such as roach, tench, and minnows there is a series of bones or *Weberian ossicles* which transmit these vibrations from the swim bladder to the inner ear. These small bones are derived from vertebrae and the sounds are transmitted to the sacculus and lagena. Fish with such connexions between the swim bladder and the inner ear may be trained to respond to sounds of higher frequencies than can be detected using the ear alone.

In tetrapods the lagena is specialised to detect airborne vibrations, which are usually transmitted to it from the outside by means of a bone, the *stapes* (columella auris), which is homologous with the *hyomandibula* of fishes. The external tympanic membrane vibrates in response to the vibrations and this bone transmits them across the middle ear to the *fenestra ovalis*, which communicates with the fluid (perilymph) surrounding the inner ear. Frogs have been conditioned to sounds of 50–10,000 cycles/sec, but they cannot distinguish the different pitch or frequency of these tones. In the reptiles an extension of the sacculus forms the beginnings of a cochlea but there is little evidence for pitch discrimination. Despite the small increase in complexity of the labyrinth, birds have a considerable range of pitch discrimination which may extend as high as 25,000 cycles/sec in the pigeon. It is in the mammals, however, that the labyrinth has become modified to such a great extent by the elongation of the cochlea and its coiled structure (Fig. 108).

8.342 *The mammalian ear*. Essentially, the mammalian ear consists of (*a*) an outer portion which functions as a sort of trumpet concerned with the collection of the airborne vibrations, (*b*) the middle ear in which lie the three auditory ossicles serving to transmit these vibrations to the fenestra ovalis, (*c*) finally the internal labyrinth with its elongated cochlea, which is the site of the nervous discrimination of the different sounds that have been transmitted to it.

The external ear or *pinna*, with its direction-finding capacity, is well developed in some mammals such as bats which have very acute hearing, and they are sometimes moved reflexly in order to facilitate the collection of sound from a particular direction. The three ear ossicles lie in the middle ear which communicates with the pharynx by the *Eustachian tube* which is derived from the hyoid gill slit. This tube is normally closed by a muscle but opens during swallowing to allow the pressure on the two sides of the *tympanic membrane* or ear drum to be equalised. The three ossicles function not only to transmit vibrations of the tympanic membrane but also to concentrate the pressure change on the oval window with which the inner ossicle (stapes) communicates. In man this bone only weighs 1·2 mg and the fenestra ovalis on which it acts like a piston has an area of 3·2 mm^2, which is about $\frac{1}{22}$ the area of the tympanic membrane. Consequently the ossicles act like a hydraulic press which steps up the force of the vibrations twenty-two fold, although their amplitude is correspondingly reduced. In this way the ear absorbs the major part of the sound energy which

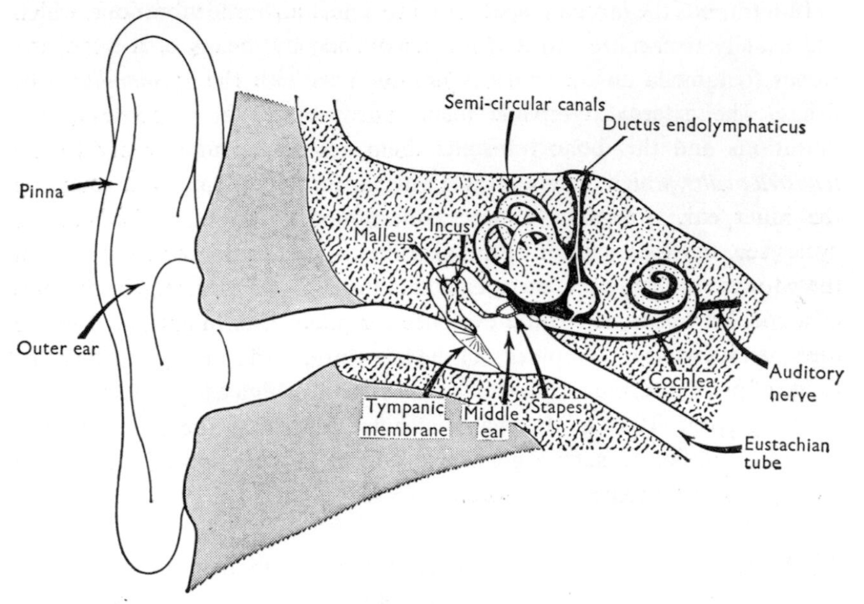

Fig. 111. The human ear.

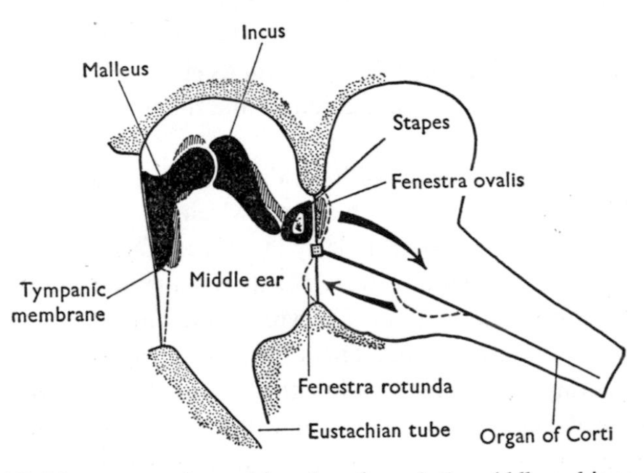

Fig. 112. The passage of sound impulses through the middle and inner ear.

impinges upon the tympanic membrane and transmits it to the inner ear. There are two muscles in the middle ear which may contract by means of reflexes when very intense sounds fall upon the ear. By increasing the stiffness of the transmission system they decrease its sensitivity and thus protect the internal mechanisms from overstimulation and are analogous in function to the iris of the eye.

We must now consider in detail the structure and function of the *cochlea*. The cochlear duct (or *scala media*) is an extension of the internal labyrinth which is attached to one wall of the bony tube in which it lies. As with other parts of the labyrinth it contains endolymph, whereas the remainder of the tube is perilymph. This perilymphatic space is divided by a partition (*the basilar membrane*) into a lower *scala tympani* and an upper *scala vestibuli* (Fig. 113). Vibration of the oval window is transmitted not to the cochlear duct or scala media itself but to the scala vestibuli. Consequently, when sounds are transmitted from the stapes the whole of the fluid within the scala vestibuli is set into vibration. Close to the oval window is the round

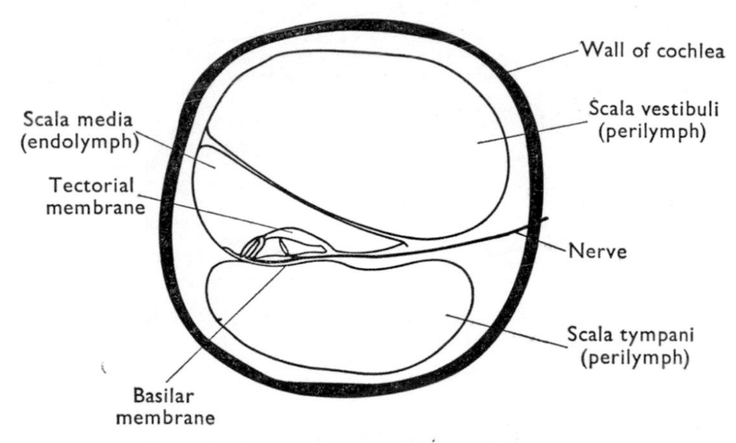

Fig. 113. Transverse section of part of the cochlea.

window or *fenestra rotunda* with which the scala tympani communicates. The scala tympani and scala vestibuli only communicate through a very fine pore (*helicotrema*) through which vibrations in the auditory range cannot be transmitted. Consequently, changes in pressure of the scala vestibuli produced by inward movements of the stapes result in pressure increases which can only be transmitted to the scala tympani by movement of the *spiral lamina.* The scala tympani connects with the round window or fenestra rotunda, which will move outwards. The basic mechanism is illustrated in Fig. 112 which indicates the condition of the cochlea when it is drawn out from a spiral condition. The importance of the spiral lamina is clearly that it will vibrate at the same frequency as the waves transmitted by the ear ossicles from the tympanic membrane. Within it lie structures which are tuned to particular frequencies and also receptors to detect the movements.

The most important part of the spiral lamina is the *basilar membrane* which is composed of 20,000 or more fibres which project from the central part of the cochlea to the outer regions. These fibres are stiff and elastic but free to vibrate like reeds and their length increases progressively from the base of the cochlea to the helicotrema. They are about 0·5 mm in length at their longest and one-twelfth of this in the basal regions. The actual receptor cells are part of the cochlear duct or scala media and they are found in the *organ of Corti.* This is composed of many hair

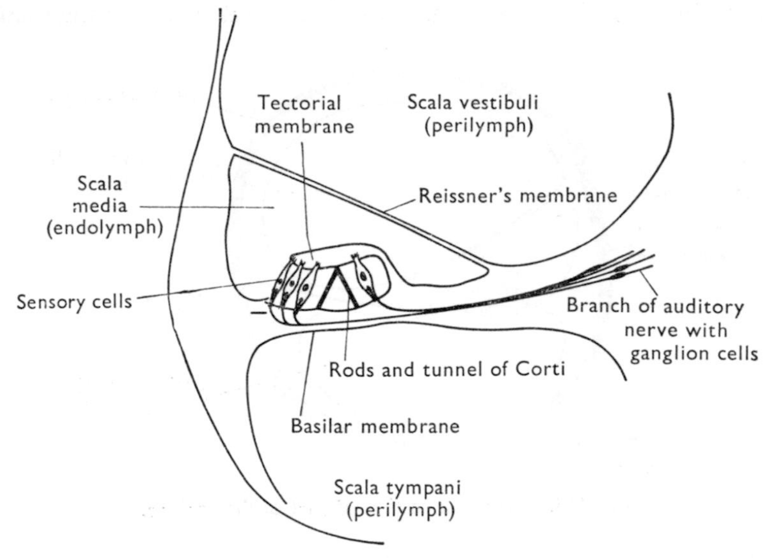

Fig. 114. Organ of Corti in enlarged part of transverse section of cochlea.

cells innervated by nerve endings from the cochlear nerve. Excitation of the endings results from the hair projections touching the *tectorial membrane* which lies above them in the scala media. When the basilar membrane vibrates, the hair cells are excited because they move upward and inwards. This shearing effect is due to the presence of a supporting structure called the rods of Corti (Fig. 114). This arrangement leads to an amplification of the pressure changes in the fluids and a greater shearing force on the sensory cells.

Vibrations of the stapes transmitted to the scala vestibuli produce deformations of the fibres of the basilar membrane. Because of their elasticity they bend towards the round window and initiate the passage of a wave down the basilar membrane. This is similar to the way in which a wave travels down a water hose when the end is moved up and down very rapidly. The pattern of these waves varies according to their frequency. At low frequencies (up to 60 cycles/sec) this vibration produces volleys of nervous impulses synchronous with each wave in the auditory nerve fibres. As the intensity of the sound is increased the number of spikes in each group increases and hence the pattern of impulses gives information about both the loudness and pitch of the sound. Above 60 cycles, however, vibrations of the basilar membrane are unequal along its length and are maximal in certain regions because of the elastic properties of the basilar fibres. Hence there is a gradual change in the mechanism of pitch discrimination, from one due to the frequency of the groups of spikes synchronous

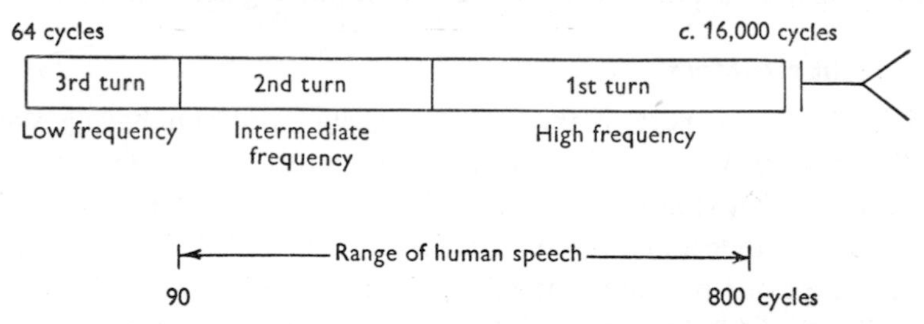

Fig. 115. The cochlea has the most exact discrimination of frequency or pitch over the ranges covered by human speech.

with the low-frequency sounds to a system which depends on the localisation along the organ of Corti of the receptors which are maximally excited. This latter system is solely responsible for pitch discrimination about 4000 cycles/sec. Because of the difference in tension in the basilar fibres and the mass of fluid which needs to be moved, the high-frequency sounds resonate at the base of the cochlea where the fibres are short and the mass of fluid to be moved is slight. Low-frequency sounds travel a longer distance in the cochlea, almost to the opposite end before bulging the basilar fibres, and these longer fibres resonate at lower frequencies.

Loudness of the sound is discriminated by the intensity of movements of the basilar fibres. More cells are stimulated and discharge at higher frequencies when these movements are greater. Not all of the analysis of sound is done by the peripheral sensory mechanism, although this is a major part of it. The auditory parts of the C.N.S. sharpen the peripheral analysis in ways which are as yet little understood.

The sensitivity of the ear is quite fantastic for at some frequencies the vibrations of the basilar membrane are only one billionth of a centimetre, that is, about $\frac{1}{10}$ the diameter of a hydrogen atom. The human ear is less sensitive at low frequencies, being only $\frac{1}{1000}$ as sensitive to tones of 100 cycles as it is at 1000 cycles/sec. At higher frequencies the human ear, in children, may detect frequencies as high as 40,000/sec, but this gradually declines with age. It is well known for example that children may hear the high-frequency squeaks of bats, but these are inaudible to most adults. In some bats the highest frequency component of their sound pulses may be 70–80 kcs. Bats are known to be able to detect objects in their path by detecting the reflected pulses of sound and making some central calculation of the delay between emission and reception. This form of echolocation may be extremely efficient and enables some bats to locate small flying insects and others to detect fish swimming beneath the surface.

8.35 *Light receptors*

Light forms the part of the electromagnetic spectrum having wavelengths between 2000 and 100,000 Å, but within this range human eyes are only sensitive between 4000 and 7500 Å (visible light). Nearly all living organisms can detect and respond to light stimuli and their receptors are able to detect one or more of the following: (*a*) the intensity of the light, (*b*) its direction, (*c*) the pattern of stimulation and the formation of images, (*d*) the frequency of the light wavelengths (i.e. colour), (*e*) the plane of polarisation of the light.

Among the vertebrates the form of the eye is relatively constant, although it is modified a great deal in relation to the particular habitat and mode of life of the organism. In the majority of vertebrates it is the principal sense organ of the whole body. Each eye arises as an outpushing of the diencephalon which, as it reaches the surface ectoderm, induces a local thickening which sinks in to form the *lens* of the adult eye. The stalk of the original outpushing forms the *optic nerve*. Because of its mode of development, the retina is inverted, that is, the light must pass through the nerve fibres before it impinges on the photosensitive elements. As with other receptors the light reception results in the passage of nervous impulses along nerve fibres to the C.N.S. These impulses are generated in the endings of the retina at the back of the eye and it is their frequency which conveys information relating to the light intensity. The change from the light energy into electrical energy requires the presence of mechanisms for the absorption of the light, and this is the function of specific pigments contained in the sensitive rods and cones of the retina. These are highly specialised cells but recent electron-microscope pictures have revealed that the outer part of the rod appears to be a very complex cilium, for it is connected to the inner region by a thin stalk, which in cross-section has the characteristic nine pairs of peripheral filaments. The outer segment is a laminated structure and the many discs contain the photosensitive pigment molecules (*rhodopsin*). The basal part of the rod contains many mitochondria. It is likely that the original photochemical reactions occur in the distal part and give rise to the release of energy in the inner part leading to the production of potential changes which set off the electrical events.

8.351 *The photochemical reactions.* As we have mentioned, the photosensitive pigment of the rods is called *rhodopsin* or *visual purple.* Light energy changes this into an unstable compound which in turn automatically gives rise to a protein component and retinene. As the latter is an aldehyde of vitamin A, deficiency of this vitamin reduces the ability to detect light, particularly at low intensity. In light the rhodopsin is being continually reformed, though not nearly so rapidly as it is destroyed because this reaction has a half-time of only a few minutes. Exactly how the breakdown reactions elicit nervous impulses in the rods is not known. The excitatory effects take place during about $\frac{1}{10}$ sec, following a sharp and instantaneous flash. It should be noticed that, even if this flash lasts one millionth of a second, the sensation is exactly the same as if it were $\frac{1}{10}$ sec because it is during this time that excitation occurs regardless of the duration of the flash.

Cones contain different light-sensitive pigments of which there are probably three in man. As with other light-sensitive pigments in the animal kingdom it is composed of retinene and a protein, but the latter differs from that of other pigments. This pigment is not so readily affected by the light and they are therefore more able to function efficiently at higher light intensities. The presence of both rods and cones in different proportions enables vertebrates to function efficiently over a wide range of light intensities in their natural habitat. Diurnal animals generally have a larger number of cones relative to rods, whereas the number of rods is higher in nocturnal animals. Apart from such differences in proportion there are also adaptations whereby the rods and cones can be moved into a protective pigment layer in intense light and they may move relative to one another. In other cases the pigment layers may be mobile.

8.352 *Acuity of vision.* Once the rods and cones are excited the nervous impulses pass through a network of fibres and across several synapses before they reach the nerve fibres contained in the optic nerve entering the brain. In some parts of the retina (fovea) each photosensitive element makes connection with a single chain of neurones along a sort of 'private' line. The information transmitted is therefore very detailed and confers a high degree of acuity on the visual system. From other parts of the retina, however, several visual elements make contact with a given intermediate neurone and consequently the total acuity is reduced. Moreover the excitation of a single rod or cone may result in impulses passing up several nerve fibres and these same nerve fibres may be excited by other rods. By and large it is only the cones that have private lines to the brain, and the rods

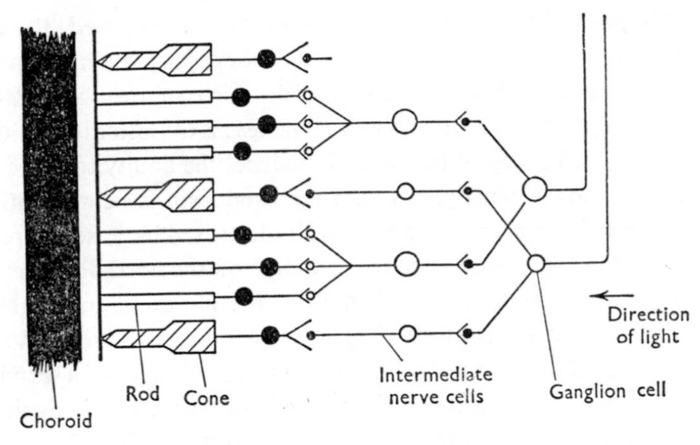

Fig. 116. The retina of the eye, showing convergent connections of rods and cones.

have more diffuse connections. It must be remembered that before the light can reach the rods and cones it must pass through all the other layers of the retina. The efficiency of the mechanism is also enhanced by the presence of a dark pigment layer at the back of the retina which prevents the reflection of any of the light after it has passed through the retinal sensory system. The absence of this dark pigment (which corresponds to the black inner surface of a camera) leads to a marked fall in the visual acuity such as occurs in albinos who hereditarily have no melanin pigment in any part of their bodies. In nocturnal animals this black pigment layer is replaced by a reflecting layer on the choroid called the *tapetum*. In this way the sensitivity of the eye to light is increased because the receptive elements will be stimulated as the light passes through to the tapetum and when it is reflected back again. As we have suggested above, however, there will certainly be a loss of acuity. Light reflected out through the pupil produces the bright shining eyes of mammals such as cats. In some dog-fishes and sharks the tapetum is formed of a layer of cells filled with guanin granules and these may be covered or uncovered by a movement of the pigment cells, according to the light intensity. This device has been evolved many times independently; for instance tapeta are also found in the eyes of seals, whales, and various species of fish, carnivores, etc.

8.353 *The eye as an optical instrument and its aberrations.* As an optical instrument the vertebrate eye may be compared to a camera. Light enters through the pupil and lens, where it becomes diffracted and is focused on the sensitive retina, where the photochemical reactions take place and the

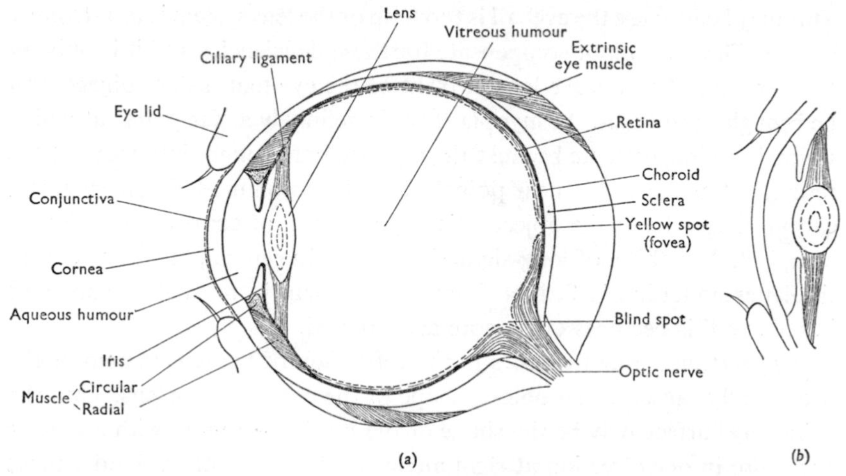

Fig. 117. The eye. (*a*) Eye in normal distant focus. The sclera pulls out the lens, giving long focal length. (*b*) Eye in near focus. The ciliary muscles draw the sides of the sclera together taking the strain off the lens, which becomes more spherical.

sensory fibres are stimulated. A great deal of refraction takes place at the air-corneal interface because of the change in refractive index but there is also a considerable change at the lens itself. For purposes of discussion it is convenient to think of all these refractive regions as one and in this case the eye could be represented schematically as a *reduced eye*. The single lens of such an eye is about 17 mm in front of the retina, on which it produces an inverted image. Nevertheless the brain perceives objects with their correct orientation because of its training to recognise an inverted image as the normal one. The shape of the lens itself may be changed from a moderately convex one to a very convex lens, and this is known as *accommodation*. In the normal eye with all parts relaxed the parallel rays from a distant object are focused on the retina, but in certain conditions these rays may become focused either behind the retina (far-sightedness) or in front of the retina (near-sightedness or *myopia*). In far-sightedness when the eye accommodates and makes the lens more convex, a distant object can be brought to focus on the retina. So long as the muscles concerned in accommodation do not have to contract too strongly, the lens is still able to become even more convex and so enable the individual to focus on nearer objects. Where the far-sightedness is extreme, however, the parallel rays may only be brought to focus on the retina by the insertion of another convex lens in a pair of glasses. In myopia the eye even when completely relaxed is unable to focus distant objects at the retina. This may be because the eyeball is too long or the lens system is too strongly convex. The eye cannot compensate for these deficiencies and it is only by the insertion of a concave lens in front of the eye that distant objects can be brought into focus. In myopia there is definitely a 'far point' at which vision is no longer acute because the rays are not brought into focus at the retina, just as there is a 'near point' when the accommodating mechanism can no longer focus the object. Near points are, of course, found for all eyes, including those of long-sighted persons. In general, the near point in the latter condition is farther from the eye than in normal persons and in old age this becomes even more accentuated.

Astigmatism is due to errors in the refraction of the lens system of the eye, usually caused by an oblong shape of the cornea or the lens. Thus the refracting surface may be the shape of the bowl of a spoon, with a greater curvature in one direction at right angles to that in another. Under these conditions objects in one of these planes may be brought normally to focus on the retina whereas the other would be focused in front of the retina. In fact all combinations are possible, and one of them is called *mixastigmatism*, where distant objects in one plane are focused behind the retina and in the plane at right angles they are focused in front of the retina. Correction for astigmatism of this common type can be achieved by the use of a cylindrical lens so that the refractive power will be altered in one direction and not in the direction at right angles to it. Spherical lenses in front of an astigmatic eye can bring the rays that pass through one plane into focus on the retina, but can never bring all the light rays into focus at the same time.

8.4 The central nervous system

As described above, the central nervous system consists of a great mass of nervous tissue lying between the receptor and effector organs. These constitute respectively the input and output elements of the system, and the C.N.S. serves to modify the relationship between input and output. The C.N.S. develops as a dorsal tube which is originally undifferentiated along its length. However, because of the modification of the head region for feeding and the presence of the major sense organs, the anterior region becomes differentiated as a *brain* and the remainder forms the *spinal cord*.

The arrangement of the neuronal pathways within the spinal cord is less complex than that of the brain region and it retains more of the segmented condition of ancestral chordates. In the brain many complex 'supra-segmental' systems are superimposed on the basic pattern. It is thus easier to start with the arrangement of the neurones and their pathways within the spinal cord before considering the complex brain.

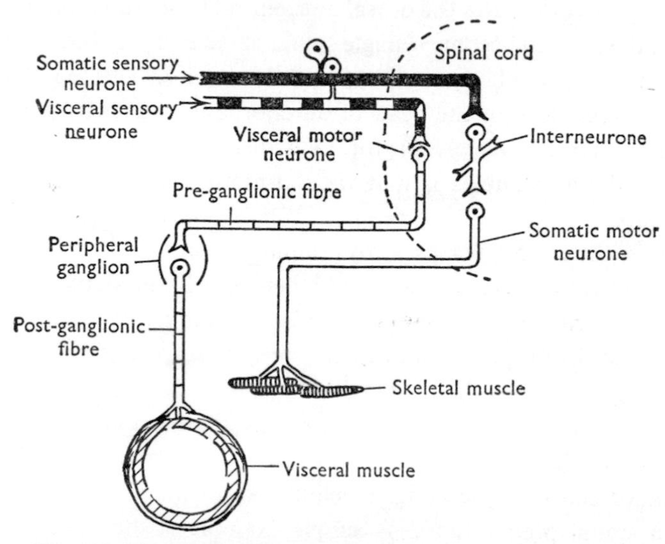

Fig. 118. The connexions of the somatic and visceral neurones through the spinal cord.

8.41 *The spinal cord*

8.411 *Functional components and the anatomical basis of the reflex* (see Fig. 118). In transverse section there is a small central canal which contains *cerebro-spinal fluid*. The central part of the cord is the grey matter, composed of cell bodies and their dendrites, while the outer region is the white matter and consists of axons passing up and down the cord. Some of these run for short distances but others form recognisable tracts descending from the brain and carrying information to the spinal nerve centres. Other tracts of ascending axons also exist which transmit impulses to the appropriate regions of the brain.

From each segment a ventral root passes out which is made up of motor fibres carrying impulses to the skeletal musculature (the *somatic motor* component) and to the gut and other derivatives of the lateral plate (the *visceral-motor* component). In the nerve cord and ventral root the visceral fibres are placed dorsally to the somatic ones. In the dorsal region of the cord the segmented dorsal roots carry sensory fibres which bring information from the skin and proprioceptive organs of the skeletal muscles (the *somatic sensory* component) and others which transmit sensory impulses from the smooth muscles and gut (the *visceral-sensory* component). The dorsal roots of some lower vertebrates also contain viscero-motor fibres but these are entirely in the ventral roots of the higher groups.

In all higher vertebrates the dorsal and ventral roots soon join after their exit from the cord to become a single spinal nerve carrying both motor and sensory fibres. In the body region of some lower chordates (e.g. Lampreys and *Amphioxus*) and in the head of all chordates, the two roots remain separate—either completely or for long distances. A further point of interest in these examples is that visceral-motor fibres are found in the dorsal roots.

There are thus two reflex pathways running within the spinal cord—that concerned with the skeletal muscles and their coordination, the *somatic arc*, and that concerned with coordination of smooth muscles, etc., the *visceral arc*. The latter forms part of the autonomic system (p. 210).

8.412 *Spinal reflexes.* The main functions of the spinal cord can be illustrated by reference to the reflexes which exist when the cord is severed from higher centres by cutting it behind the medulla of the hind brain (i.e. in a spinal preparation). A simple example is the so-called *stretch reflex* by which muscles that have been stretched contract more strongly because of excitation through a spinal path. An everyday example is the reflex producing the *knee jerk*, when the tendons of the knee cap are stretched and thus result in excitation of the knee by contraction of these muscles whose tendon has been stretched. In such a reflex only two neurones need be involved, the afferent (sensory) neurone conveying impulses from the stretch receptors, which are the muscle spindles in the body of the muscle, and the efferent (motor) neurone which produces the muscular contraction.

Another spinal reflex is the *flexion reflex*. If the toe of a spinal cat, that is, one with the spinal cord sectioned behind the medulla, is pinched, its limb will be flexed away from the point of stimulation. At the same time as it lifts one limb, it can be seen that the other one of the pair on the opposite side is extended—a natural reflex which would stop it falling

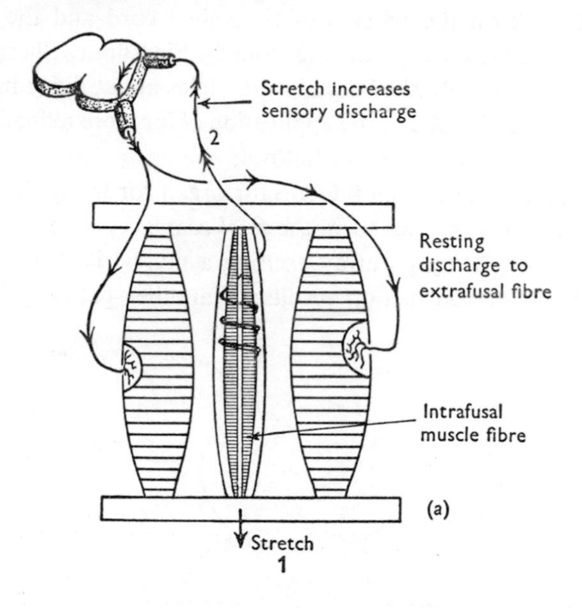

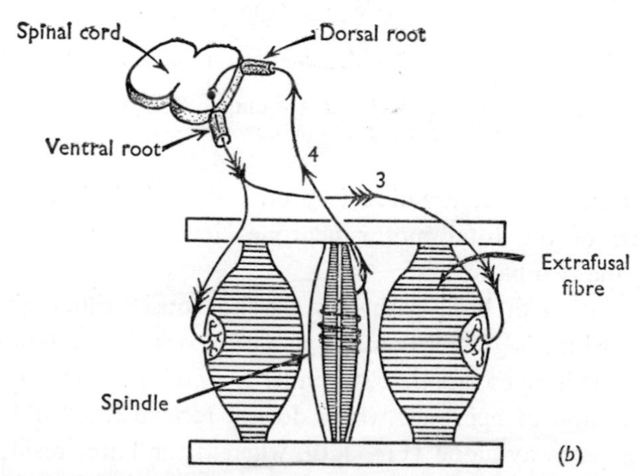

Fig. 119. Stretch reflex. (*a*) Stretch; (*b*) result. The numbers indicate the sequence of events and the frequency of nerve impulses is indicated by the arrows.

over—and this is the *crossed extension reflex*. From detailed investigations of the flexion reflex by micro-electrode techniques, it appears that at least three neurones are involved—the afferent, interneurone (within the spinal cord) and efferent. This is an unusually simple example and it must be realised that normal reflex pathways are often more complex, especially with regard to the number of interneurones that are involved.

Classical work on the reflexes of the spinal cord and the integrative properties of the nervous system was done by Sir Charles Sherrington and his associates who established certain principles about reflex mechanisms. These are still valid in spite of the application of far more refined techniques and some may be summarised as follows:

(*a*) A single muscle can be the effector organ for many different reflex arcs (i.e. its motor neurones form the *final common path*).

(*b*) If a given sensory pathway exciting a muscle is stimulated then a certain strength of contraction results; if another pathway is excited a

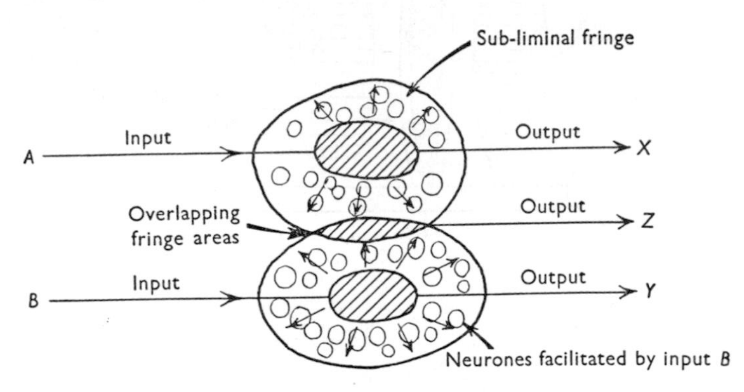

Fig. 120. Spatial summation. Input $A \rightarrow$ output X; input $B \rightarrow$ output Y; but input $A+B \rightarrow$ output $X+Y+Z$.

different strength of contraction is produced. This is because different proportions of the total motor neurones innervating the muscle are excited by the two pathways.

(*c*) Stimulation of both pathways simultaneously either gives a total contraction which is greater than the sum of separate contractions (*spatial summation*) or it gives one that is smaller (*occlusion*). The former is due to the excitation of neurones which do not receive threshold excitation from either pathway alone (Fig. 120), whereas the latter results from an overlap of the neurones which were excited supra-threshold by both pathways and therefore summation is not possible.

The above shows in itself how complicated can be the performance of a given effector system following stimulation through spinal reflex arcs.

8.413 *Spinal reflexes, posture, and locomotion.* (*a*) *Tetrapod locomotion.* By analysing the movements of a four-legged animal, such as a dog, when it is walking slowly it is possible to see how a combination of the spinal reflexes is involved in producing the normal walking rhythm. In this rhythm the legs move in the order right fore, left hind, left fore, right hind, right fore, etc. Thus when one leg is lifted the other leg of a pair must support a greater proportion of the body weight, as must also the limb on the same side. In order for this to operate, the stretch reflex is called into action; consequently the extensor muscles of the three legs on which the dog is standing increase their activity in proportion to the weight they support. Besides the stretch reflex the crossed extensor reflex also increases the extensor action of the contracted limb and a further type called the ipsilateral extensor reflex increases the tension in the extensor muscles of the leg on the same side. All three reflexes are operating in the normal walking of the dog.

Of vital importance in all of these reflex responses is the inhibition between the muscles which operate antagonistically across a given joint. Stimuli which result in the excitation of a given flexor muscle have the effect of inhibiting activity in the extensor motor neurones. The basis of this reciprocal inhibition of antagonistic muscles depends on the different effects on the motor neurones of the excitatory and inhibitory inputs to the C.N.S. The former tend to depolarise the soma membrane and thus tend to produce excitation, whereas inhibitory pathways increase the polarisation and therefore hinder the production of a propagated impulse.

(*b*) *Swimming*. Just as the locomotory rhythms of a tetrapod can be interpreted in terms of spinal reflexes, so the swimming movements of fishes involve reflexes of both segmental and intersegmental types. If in a dogfish the spinal cord is cut behind the medulla, the fish shows swimming movements which may persist for many days without stopping. These movements may be inhibited if the fish touches the bottom of the tank and are increased in amplitude by putting a clip on the tail—clearly showing that sensory input from the skin affects these movements. The receptors involved in the coordination of the normal rhythm are more likely to be proprioceptors in the muscles or connective tissue, and if all sensory input to the cord is removed by cutting the sensory nerves (i.e. cutting all the dorsal roots) then the spontaneous swimming movements cease.

From experiments of the above type on vertebrate locomotion we are led to the conclusion that the spinal cord contains all the reflex pathways which can interact with each other to produce normal locomotion. Of course these reflex paths are not independent of the higher centres in the intact animal and for finer gradations of movement these higher centres are necessary as well as being needed to assess the appropriateness of responses and in many cases to initiate them.

8.414 *The autonomic nervous system and visceral reflexes.* The motor component of visceral reflex arcs, whose path through the spinal cord has already been described, makes up the autonomic system. This system

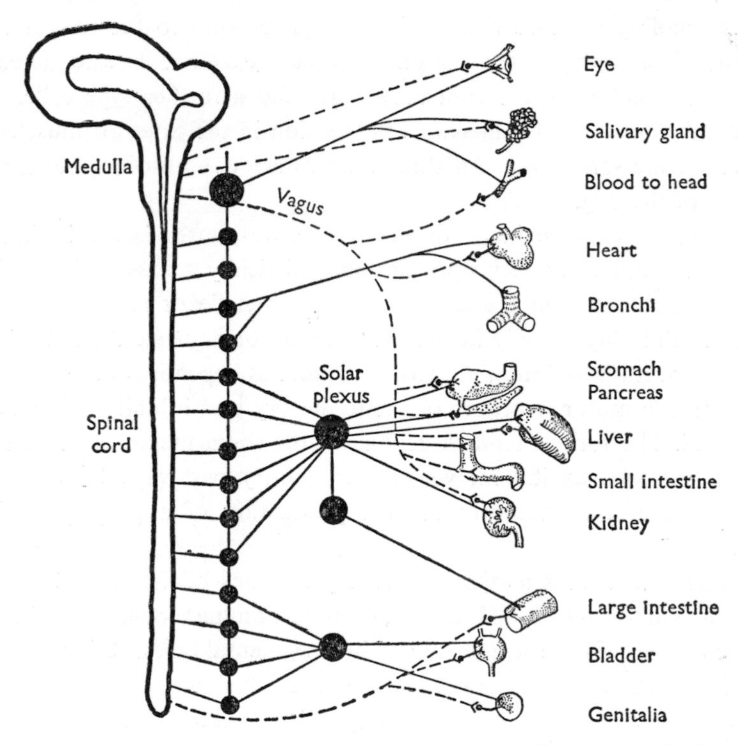

Fig. 121. The autonomic nervous system. —— Sympathetic; ●, sympathetic ganglia; - - -, parasympathetic; - -<·, parasympathetic ganglia.

coordinates the activities of the viscera which take place below the conscious level and it can be subdivided into the *sympathetic* and *parasympathetic.*

In the mammals the distinction between the two systems is clear both structurally and functionally. In both there is a ganglion outside the central nervous system where a pre-ganglionic neurone (i.e. one leading from C.N.S. to the ganglion) synapses with a non-medullated post-ganglionic neurone; that is, one leading from the ganglion to the organ innervated, such as the intestine, heart, etc.[1] There is, however, an anatomical distinction between the sympathetic and parasympathetic systems in the relative lengths of the pre- and post-ganglionic fibres. In the sympathetic system the ganglion is near the C.N.S. but in the parasympathetic—for example, the vagus—the ganglion is located near the organ to be innervated, such as the heart. Some of the ganglia of the sympathetic system are in an approximately half-way position. One of the results of this arrangement is that the effects of parasympathetic stimulation tend to be very localised, while sympathetic excitation spreads more widely over the area innervated. This spread is also a feature of the grosser structural arrangements, as the outflow of the sympathetic system extends throughout the thoracic and lumbar regions of the spinal cord, whereas the parasympathetic fibres have their exits only in the cranial and sacral regions of the C.N.S. The spread of sympathetic stimulation is further encouraged by the passage of some preganglionic neurones up and down the sympathetic chain which joins some of the sympathetic ganglia.

Functionally the two systems differ in the nature of the transmitter substance released at the endings of the post-ganglionic neurone. In both parasympathetic and sympathetic ganglia the pre-ganglionic fibre liberates acetyl choline which excites the post-ganglionic neurone. At its endings on plain muscle or a gland the latter neurone produces acetyl choline in the case of parasympathetic neurones but in sympathetic systems the post-ganglionic neurone liberates the adrenaline-like substance, sympathin or noradrenaline. Sympathetic fibres to sweat glands in man are cholinergic. The adrenal gland itself is a sort of giant post-ganglionic neuronal system liberating vast quantities of adrenaline into the blood system and it generally acts in conjunction with sympathetic stimulation.

The most important difference between the two systems, which is clearly related to their structure, is that the parasympathetic innervation is more localised and produces more discrete responses than does the sympathetic system. The sympathetic generally functions as a whole and its effects may be summarised by considering the reactions of the body which take place in a state of emergency. Under these conditions the heart rate is increased, the blood pressure is raised, the respiratory rate increases and blood is removed from the gut and tends to flow more to the muscles. All these effects are the result of sympathetic stimulation. It is true that the para-

[1] The visceral-motor components of the cranial nerves which supply the muscles of lateral plate origin operating the visceral arches have no ganglia, do not form part of the autonomic system and are sometimes distinguished as the special visceral-motor component.

sympathetic generally produces the opposite effects, for example, slowing of the heart, constriction of the blood vessels to the muscles, but there are some instances in which both systems have apparently identical effects but play a part at different times. Thus, the secretion from the salivary glands differs according to whether it is produced by parasympathetic or sympathetic stimulation. Similarly, it is not valid to say that adrenaline always produces the same reaction as does sympathetic excitation. For example, there is no sweat secretion when the adrenaline concentration of

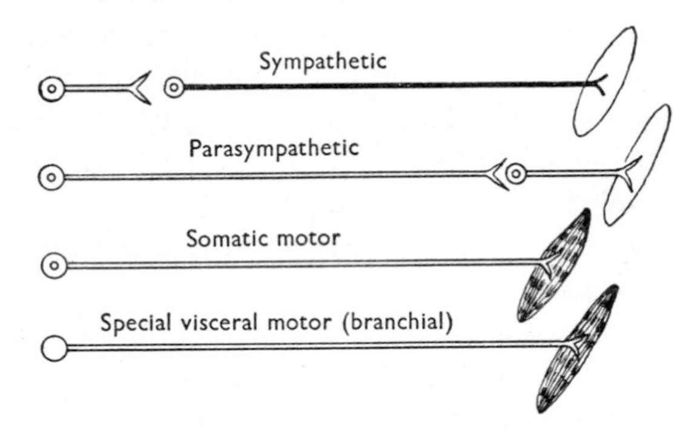

Fig. 122. Major differences between the various types of nerve–muscle systems. —— Adrenergic; ═══ cholinergic.

the blood increases, although these glands have a distinct sympathetic supply but no parasympathetic innervation. Blood contains marked quantities of enzymes which destroy acetyl choline, so that its effects are inevitably localised to the regions where this transmitter is released.

The afferent fibres (visceral-sensory) are contained in the autonomic nerves and are usually non-myelinated. They come from receptors in the viscera and other parts of the body and pass uninterrupted through the peripheral ganglia of the sympathetic and parasympathetic systems. Their cell bodies lie in the dorsal root ganglia and on entering the C.N.S. they branch profusely and may have effects on many widespread parts of the body; for example, stretching of the intestine may produce dilatation of the pupil! Many of these visceral afferents perform vital parts of the functioning of other physiological systems such as respiration and circulation. Notable examples are the afferent fibres which arise from the carotid sinus and carotid body and enter the brain stem in the ninth cranial nerve. These receptors detect respectively stretching of the carotid artery and changes in the O_2 and CO_2 content of the blood. The integration of autonomic functions is maintained by specific paths of the C.N.S., and in the brain the hypothalamus is of very great importance, as will be discussed later (p. 222).

8.42 *Brain*

The brain forms as a series of enlargements of the anterior end of the neural tube. These enlargements develop in relation to the main sense organs which, as in all bilaterally symmetrical animals, become aggregated at the front of the animal. It is interesting to notice, for example, that in the primitive chordate *Amphioxus*, which has relatively few sense organs anteriorly, there is only a slight enlargement of the neural tube to form a brain. The enlargement is mainly due to an increase in the number of nerve cells but also to the greater number of nerve fibres which enter this part of the brain from the distance or teloreceptors. The regions where these nerve endings are to be found form the *primary sensory centres* of the brain where the initial sorting out of the information takes place. But by far the most important feature of the brain is the development of *association areas* or *correlation centres* where information from several different sense organs is gathered together and analysed. As a consequence of these processes, which as yet are little understood, descending motor pathways are excited and set into action movements which are related to the pattern of sensory stimulation to which the body is subjected. The association centres also function as storage elements and form the sites of memory and it is here that the possibilities of learning are mainly found.

8.421 *Development.* In early vertebrate embryos there are usually three swellings of the anterior part of the neural tube, forming the fore, mid, and hind brains. A little later the fore brain becomes constricted into two regions, an anterior end brain or *telencephalon*, and a 'tween brain or *diencephalon*. The mid brain or *mesencephalon* does not subdivide, but the hind brain forms the anterior *metencephalon* and posterior *myelencephalon* (Fig. 123).

A further process which takes place in higher vertebrates is that these regions do not retain their primitive arrangement but the axis of the brain becomes bent, typically in three places. This results in a more compact arrangement, particularly in mammals where the cervical flexure results in the axis of the brain being almost at right angles to the axis of the spinal cord. In the later development of most mammals this flexure becomes reduced but in man it persists in the adult.

The central canal of the spinal cord continues into the brain where it forms the *ventricles*, which contain cerebro-spinal fluid as does the spinal canal. This fluid is produced by the vascular *choroid plexi* which develop in the thin roof of the 'tween brain and medulla. The whole C.N.S. is covered by an inner membrane or *meninge* (the *pia mater*) which is very vascular and helps to form the choroid plexi. Outside the pia is the *arachnoid* layer forming the web-like meshwork, which is very delicate and

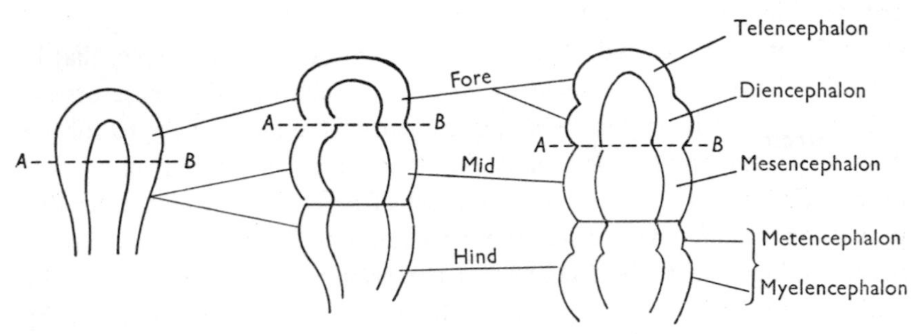

Fig. 123. Development of the vertebrate brain.

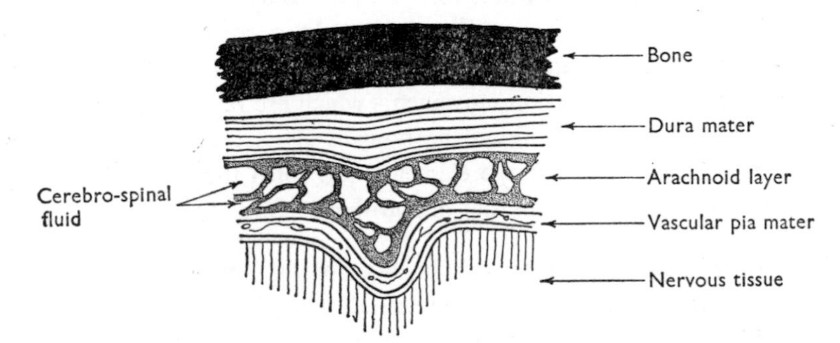

Fig. 124. The membranes covering the central nervous system.

hence its name. The cerebro-spinal fluid fills in the space between the arachnoid and pia and forms a cushion layer about the C.N.S. as well as serving to convey oxygen and nutrients to the inside when it is secreted at the choroid plexi. The outer meninge is the tough *dura mater* which protects the entire C.N.S.

8.422 *Evolution of the brain.* As we have seen, the brain of *Amphioxus* is scarcely visible, but in cyclostomes and fishes the brain is well-marked and has the same divisions as in man. In these lower forms the main subdivisions of the brain retain their primitive association with the important sense organs. Thus the telencephalon may be referred to as a 'smell brain', the mid brain as a 'sight brain', and the myelencephalon as the 'ear brain'. The relative sizes of these different portions of the brain are correlated with the degree of development of these teloreceptors and their importance in the life of the animal.

As we ascend the vertebrate series there are indications of increasing *cephalic dominance* as shown, for example, by the effects of severing the brain from the rest of the C.N.S. If the spinal cord is cut just behind the medulla in man or a mammal there is complete loss of movement,

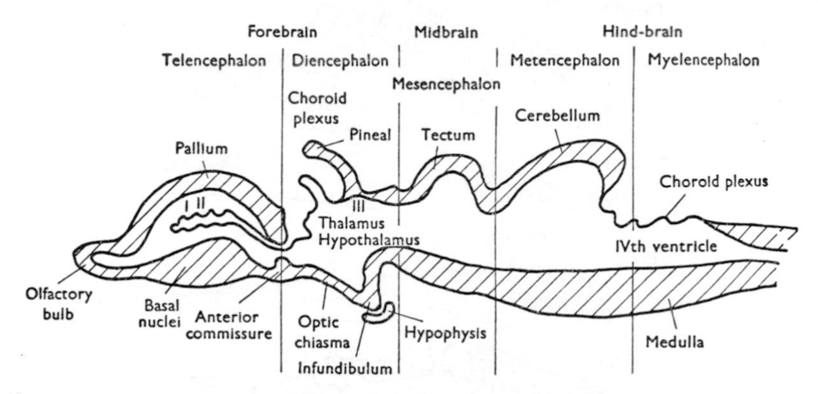

Fig. 125. Longitudinal section of a generalised vertebrate brain.

whereas, in the case of a dogfish, there persists a rhythmic swimming which may continue for several days. The hind brain remains relatively unchanged in the vertebrate series for it is mainly concerned with the control of visceral functions such as respiration and circulation. It evolves in relation to the changes in these systems but there is no fundamental change in function which might give rise to alterations in the external form. The cerebellum, which forms part of the metencephalon, does increase in birds and mammals where it plays a large part in the control of delicate movements. It is also quite large in some fishes, where movement must be oriented accurately in all three planes. The mid brain remains as a centre for optic stimuli throughout the vertebrates and in fishes the roof or *tectum* forms the most important correlation centre where tracts converge from the olfactory and vestibular parts of the brain. In fishes, amphibians and reptiles there are usually two optic lobes but in mammals these are replaced by the *corpora quadrigemina*.

It is the fore brain, however, which changes most markedly in the evolution of vertebrate brains. The fore brain of cyclostomes shows a relatively slight enlargement at the front end of the neural tube but in fishes and amphibians there is an enlargement of the roof and/or floor of this region which increases in mammals and birds. Enlargements of the roof or *pallium* lead to the great expansion of this region in the mammals so that it covers the whole of the mid brain and forms a major part of the brain (80 % by weight). The floor of the telencephalon becomes enlarged in the birds and forms the *corpus striatum*. In both cases this great development of regions within the fore brain is correlated with its increasing

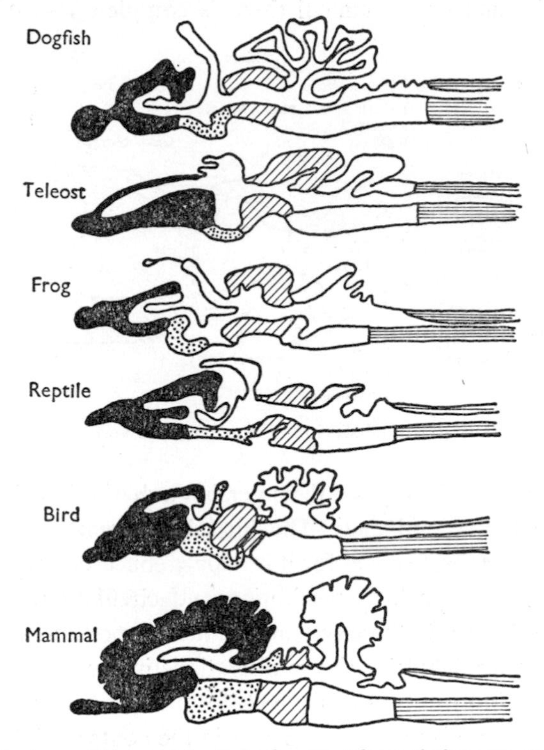

Fig. 126. Regions of the vertebrate brain. ■, telencephalon; ▦, diencephalon; ▨, mesencephalon; □, metencephalon; ⊟, myelencephalon.

importance as an association centre. In the lower forms it is concerned with olfactory sense alone but in the reptiles are found the first signs of pathways from the mid and hind brains entering the roof of the fore brain. These and other projection areas are enormously developed and have been mapped in great detail in man and other mammals. Mapping has also been done of regions concerned with motor functions by the application of small electric currents through fine electrodes. Another general feature of fore-brain evolution is the development of the *cerebral cortex*. The primitive position of the cell bodies in the C.N.S. is near the central canal, but in the brains of higher vertebrates many are also found in the surface of the brain. The number of cells in these regions is enormous and the possible ways in which the information entering through the sense organs can be sorted appear to be almost infinite.

8.423 *The main regions of the brain and their functions.* The precise position at which the brain begins and the spinal cord ends is not clearly defined for the *medulla oblongata* passes imperceptibly into the spinal cord. In general it may be taken that the brain includes that part of the neural tube which is contained within the cranium.

The *medulla*, although relatively unspecialised, forms a most vital part of the brain of vertebrates because it is concerned with the control of many essential visceral functions such as respiration, circulation, and the heart, etc., without which none of the higher functions of the brain would be possible. The medulla is the region where the majority of the cranial nerves (V–XII) take their origin and associated with these nerves are discrete groupings of nerve cells called *nuclei*. In some cases the medulla is lobed in relation to these as, for example, the vagal and facial lobes. The coordination of respiratory and cardio-vascular responses takes place as a result of the bringing together of inputs of many different sorts, but in addition neurones within the medulla have properties which enable it to produce rhythmic outputs along motor fibres resulting in the respiratory movements. The medulla of fishes and higher vertebrates can produce these bursts in the absence of any inflow from the peripheral sense organs for some time after isolation (see chapter 5). Neurones within the medulla may also respond directly to changes in the content of the blood which circulates in this region, notably it is sensitive to the CO_2 tension. Other reflexes that are mediated by the medullary nuclei include those which regulate the rate of heart beat, degree of constriction of the capillaries, salivary secretion and swallowing. The position of the nuclei and nerve terminations is basically the same as in the spinal cord; that is, somatic sensory neurones form the most dorsal part and the somatic motor components the ventral regions (Fig. 127). The cerebro-spinal fluid which forms in the anterior choroid plexus passes posteriorly and leaves the cord through three foramina in the roof of the medulla.

The VIIIth nerve, which terminates in the vestibular nucleus of the medulla, contains important afferent fibres which play a large part in the control of posture and balance in vertebrates. These afferents are able to affect directly some of the descending motor pathways responsible for producing the animal's movements. Drastic lack of input on one side results in rolling of a fish or tetrapod, though some compensation may occur. A frog, for example, may compensate after labyrinthectomy in six to eight weeks. In fish and amphibians, section of the nerve cord in front of the medulla leaves the animal still capable of relatively normal locomotion. In the dogfish even if the cord is sectioned behind the

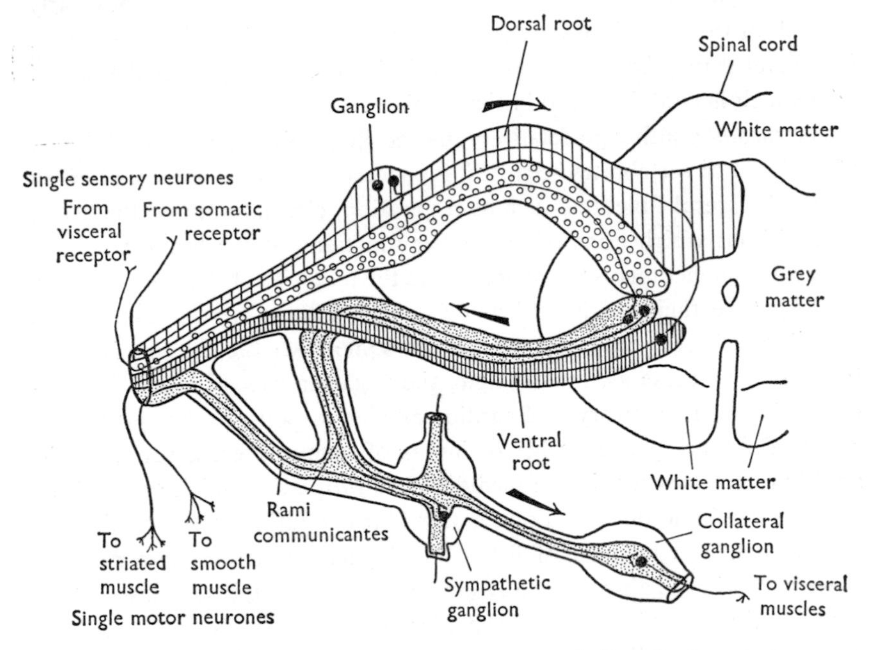

Fig. 127. The spinal arc. The major nerve tracts are indicated; these carry many individual sensory and motor neurones. [vertical lines] somatic sensory; [dense lines] somatic motor; [circles] visceral sensory; [stipple] visceral motor.

medulla the resulting spinal preparation continues to make spontaneous swimming movements for many days. In birds and mammals, however, locomotion is not possible even in a medullary animal (i.e. C.N.S. sectioned in front of the medulla) but it continues to make its basic respiratory and cardio-vascular responses.

As we have seen, the acoustico-lateralis centres form a very important part of the medulla and are situated dorsally. In this region the anterior part of the medulla develops an extension known as the *cerebellum*. In accordance with the region with which it develops, its main function is that of balance and motor coordination and it has been referred to as the 'gyroscope of the body'. It forms the *metencephalon* as distinct from the *myelencephalon* which comprises the medulla proper. The cerebellum is concerned with fine gradations of posture and orientation and becomes greatly developed in birds and mammals. Its surface is complexly folded and contains some of the cell bodies in the form of a cerebellar cortex. In

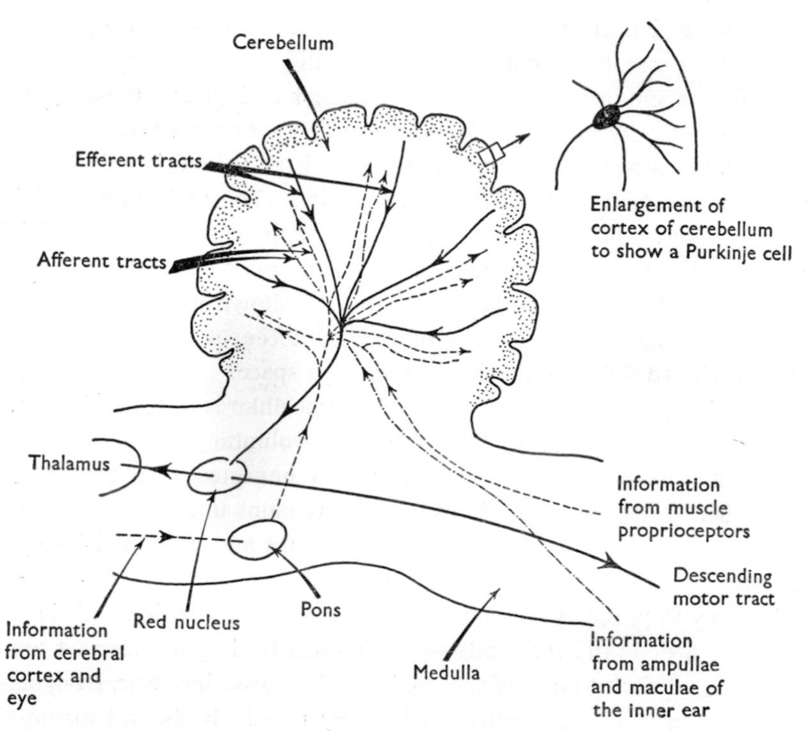

Fig. 128. The main connexions of the cerebellum (median longitudinal section).

sections of the cerebellum (Fig. 128) these folds are seen to be lined by the characteristic *Purkinje* cells. These are the only efferent cells of the cerebellar cortex and their axons terminate in some of the sub-cortical nuclei. The close connexion between the cerebellum and cerebral hemispheres is further indicated because the latter develop both *ontogenetically* (life-history of the individual) and *phylogenetically* (race-history of the species) at about the same time as the so-called *neocerebellum*. Correlated with the development of this interrelationship in mammals is found the *pons*, which contains nuclei that relay cerebral impulses to the cerebellum. The pons also contains transverse fibres interconnecting the two sides of the cerebellum. There appears to be some functional localisation within the cerebellum, for stimulation of a given region either inhibits or facilitates the effects produced by stimulation of corresponding areas of the motor cortex. The cerebellum is characterised by electrical activity of high frequency (150–250 per sec).

Removal of the cerebellum in a dog, cat or monkey gives rise to complex symptoms which are essentially the same in each animal. They become incapable of making voluntary movements and cannot stand. They make periodic, spasmodic movements of the leg, neck and tail and these extremities become rigidly extended. After a period of four or five weeks some sort of voluntary movements return in a dog and the animal is able to move about on its four legs, but very unsteadily. The cerebellum is highly developed in birds and is very important in their sense of balance. Birds cannot walk or peck, let alone fly, several days after its removal. In man indications of damage to the cerebellum are shown by the tendency to produce oscillatory movements of the hand (i.e. a tremor) when the patient tries to follow a given object through space.

By and large, then, the function of the cerebellum is concerned with the more detailed regulation of the posture and voluntary movements of the animal and in doing this it is closely interconnected with the cerebral hemispheres. Its precise mode of function remains uncertain though it is certainly particularly well developed in primates and man, which rely on their ability to manipulate objects in their environment.

In many ways the *mid brain* is the most conservative region and changes relatively little in the vertebrate series. Its relative importance as a centre correlating different types of sensory input decreases, however, from being dominant in fish to a relatively subsidiary role in birds and mammals. The roof or *tectum* receives a point-to-point projection of the retina in fishes and it also receives olfactory, acoustico-lateralis and proprioceptive inputs. In addition it is the region in lower vertebrates where the main descending somatic motor tracts have their origin. Its roof is usually formed into a pair of *optic lobes* but in mammals there are four such lobes, referred to as the *superior colliculi* and *inferior colliculi*. The former retain their optic function and are mainly concerned with the reflex control of eye movements; control of visual responses requires the visual cortex. Localised stimulation of the superior colliculi can lead to discrete movements of the eyes. The inferior colliculi are concerned with auditory functions and reflexes. For example a cat without its auditory cortex can discriminate tones and detect sounds of low intensity, but this ability is lost when the inferior colliculi are also removed. The floor of the mid brain is important because it contains a large group of fibre tracts or *crura cerebri* which connect the cord with the anterior cerebral regions. In addition to these there are many nuclei, primarily of the third and fourth cranial nerves, which control the eye muscles. Another part of this basal

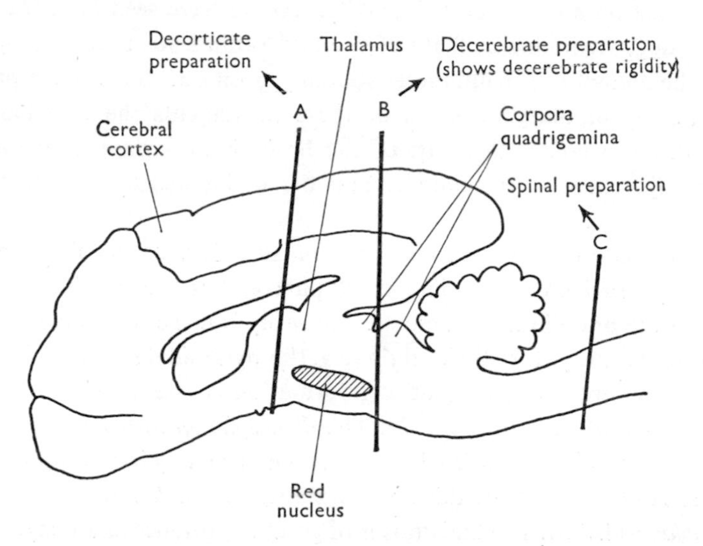

Fig. 129. Diagram showing different levels of brain section (A, B, C = cuts).

region or tegmentum forms a mass of grey matter and the so-called *reticular formation.* This is a tangled complex of branching dendrites but also contains groups of cell bodies. It appears to be important in affecting the level of excitability of other parts of the brain in a non-specific way. Appearing first of all in the reptiles, an important region of this system is differentiated as the *red nucleus* which plays an important part in the regulation of posture and movements of mammals. This nucleus is a part of the extra-pyramidal motor pathway.

Transection of the mid brain behind the red nucleus leads to exaggerated contraction of the anti-gravity muscles and produces the condition known as decerebrate rigidity. The limbs are held stiffly and point downwards and backwards and the tail and head are held high up in spite of their weight. Section of the brain in front of the mid brain of a dog or cat produces a condition in which the animal is able to right itself and stand awkwardly, but in primates standing is not possible. Decerebrate rigidity is lost when the brain stem is cut behind the vestibular nucleus and it can also be abolished in a limb by severing the dorsal roots, showing that the input from the proprioceptors is important in its maintenance. The control of the posture of the limbs is produced from the inputs derived from the muscle receptors and from the eyes. All these inputs come together in the medulla and mid brain but if the latter is removed then the postural limb reflexes take on an exaggerated activity. As we have seen in at least the cat and dog, sectioning in front of the mid brain leads to a condition in which the animal can maintain its standing position. But for the proper control and voluntary direction of these movements the fore brain is essential. In primates this part of the brain is important even for the proper adjustment of the muscles and the maintenance of posture and righting reflexes.

After considering the control of movement it is apparent that the *fore brain* plays an increasingly important part and is the dominant motor region of the brain in the mammals. Not only is it the region from which the motor pathways take their origin (e.g. the pyramidal tract) but it is also the region where a vast amount of information is assembled from inputs derived from all parts of the body. The *diencephalon* forms a smaller and more posterior part of the fore brain but contains many important regions. The walls of this region are thickened and form the *thalamus* above and the *hypothalamus* below. The thalamus is of great importance as a relay station between other regions of the brain and the cerebral cortex. Notably, for example, all optical pathways to the cerebral hemispheres pass through the thalami. In fact it is the presence of such pathways which leads to the neopallium, which first appears in reptiles and then becomes so elaborated in mammals that it covers over almost the rest of the brain (Fig. 130).

Pathways from the cerebral cortex descending to the lower regions of the C.N.S. also relay in the thalami, which is not simply a relay station on the upward and downward pathways but is involved in some complex integration with the cerebral cortex. There are several types of nuclei within the thalami and these are concerned with the afferent projection systems from the cerebellum and various other parts of the fore brain. There are scarcely any regions of the brain which are not connected with the thalami.

The ventral region of the 'tween-brain is expanded to form the hypothalamus. As with the thalami, the importance of this region of the brain has become increasingly recognised by neurophysiologists in recent years.

The hypothalamus is mainly concerned with the regulation of visceral functions, for many visceral-sensory fibres end here. Its action is not only through the autonomic system but also via the endocrine system. This is

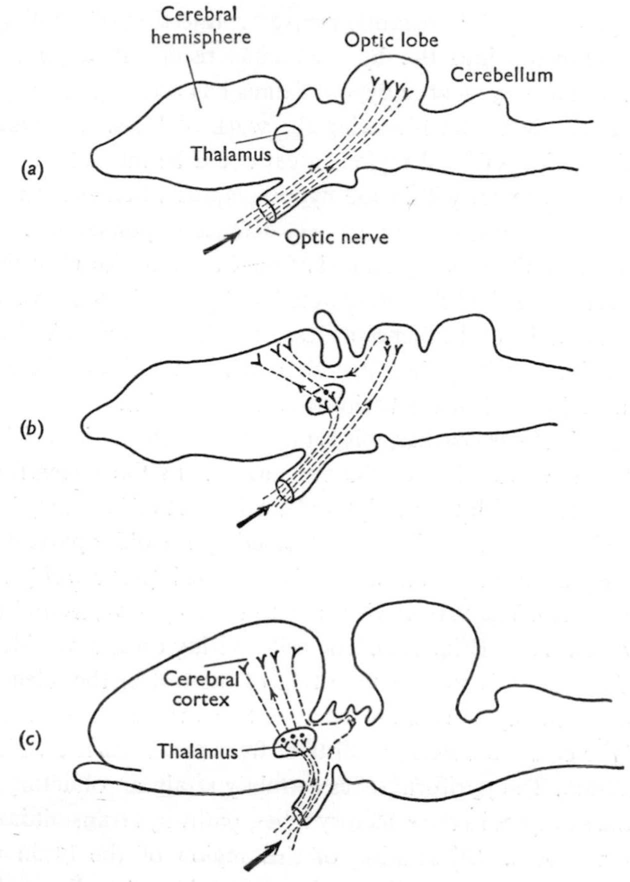

Fig. 130. The centralisation of information reception in the fore brain. (*a*) Amphibian. (*b*) Reptile. The optic lobe is reduced and information from the eyes passes to the cerebral cortex via the thalamus. (*c*) Mammal. The optic lobes are very small (represented by the corpora quadrigemina) and most of the sensory information from the optic nerves passes to the cerebral cortex (neopallium) via the thalamus.

emphasised by the presence on the base of the hypothalamus of the hypophysis, which is an extension of the infundibular stalk. Some of the hormones that are stored and released from the posterior pituitary are even produced in the hypothalamus (e.g. that regulating the urine flow). And some of these secretions pass down the axons from groups of neurosecretory cells in the hypothalamus to the pituitary (Fig. 134). The hypothalamus also contains cells which detect the body temperature and it is the site of the thermostat of homoiothermic animals. There are also osmoreceptors in the hypothalamus which form the sensory side of a neurohumoral reflex controlling the body fluid osmotic pressure. Injection of small quantities of sodium chloride into the hypothalamus results in a goat drinking profusely. In recent years the hypothalamus has been studied a great deal with implanted electrodes following the work of Hess. He first showed that by stimulation of local regions a cat could be made to sleep and to awaken. Other centres within the hypothalamus when excited may lead to vomiting, salivation, sniffing, licking, defence responses, etc.

In the roof of the diencephalon is found the anterior choroid plexus, where the cerebro-spinal fluid is secreted and passes into the ventricles of the brain. The cavity of the diencephalon is the third ventricle and anteriorly this communicates with paired lateral ventricles which are the cavities of the most anterior part of the brain—the *telencephalon.*

As we have indicated earlier, this is the region of the brain which changes most during the evolution of the vertebrates. In the lower forms it is entirely concerned with the receipt and sorting out of olfactory inputs. In the mammals this function is still performed by the older parts of the fore brain, namely the *rhinencephalon.* This is divisible into the *archipallium* and the *palaeopallium.* The former gives rise to the *hippocampus* and the latter to the *pyriform lobes* (Fig. 131). The cell layering on the outside of these regions of the brain is simpler than that of the rest of the telencephalon. They have fewer layers and are referred to as the *allocortex* (a specific region of the cerebral cortex) as distinct from the *isocortex* which covers the neopallium. The pyriform lobes certainly retain an olfactory function for after the synapses in the olfactory lobes, pathways transmit impulses to the pyriform lobes. Stimulation of this region of the brain produces actions that are related to feeding such as retraction of the lips and sniffing. Removal of the pyriform lobes leads to the loss of olfactory conditioned reflexes. The hippocampus, on the other hand, seems to have more doubtful relationships to olfactory function for stimulation and removal experiments suggest that it is more important in the production of emotional responses concerned with fear, anger and defence. It is particularly large in primates and whales but the significance of this is not known. The large size of the hippocampus in whales also contrasts with the absence of any olfactory cortex or olfactory organ. The primitiveness of these two regions of the brain is further emphasised by the fact that the input to them does not pass through the thalamus as is the case with the projection to the other parts of the telencephalon.

In birds the region of the brain which becomes most enlarged is the ventral portion of the end brain which forms the large corpora striata (Fig. 131). The neopallium is also developed in these animals but not so

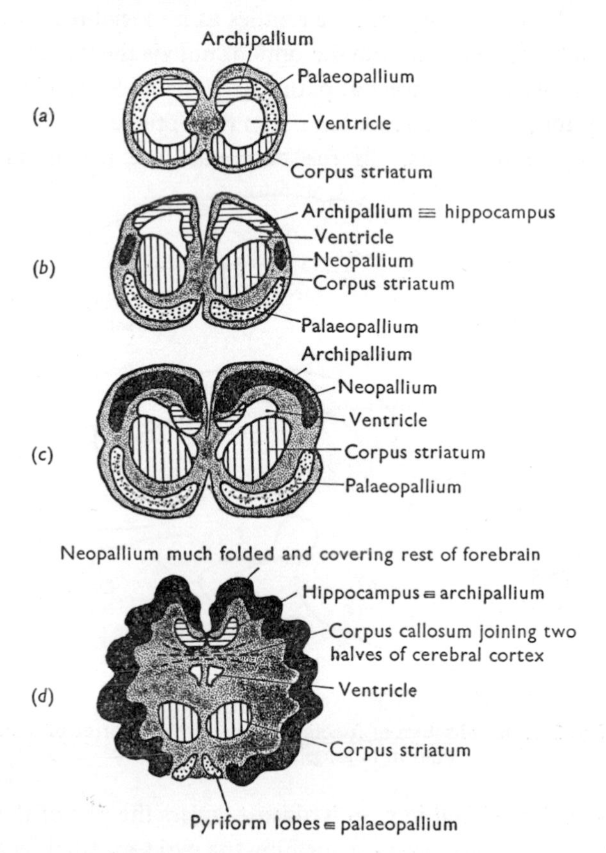

Fig. 131. Changes in the telencephalon of different vertebrates as seen in transvere section. (*a*) Amphibian; (*b*) reptile and bird; (*c*) primitive mammal (monotreme); (*d*) advanced mammal (primate).

strikingly as in the mammals. The mammals likewise have the corpus striatum which forms one of the so-called basal nuclei in the ventral parts of the telencephalon. The corpora striata function as motor centres which may be activated by impulses from the olfactory bulbs and they seem to play an important part in the control of motor activity. Together with the reticular formations of the mid brain and medulla they form an accessory motor pathway (extrapyramidal) distinct from that coming directly from the neopallium. The detailed functioning of the corpora striata is not understood but lesions in them produce tremor of voluntary movements and may result in Parkinson's disease.

But by far the most striking part of the telencephalon is the *neopallium.* This region evolved first of all in the reptiles as a special region of the fore brain, to which fibres passed from the optic input via the thalami (Fig. 130). It gradually replaced the ancestral pathway to the mid brain in controlling optic perception, although as we have seen the optic tectum still maintains control of eye movements. In the mammals the neopallium became

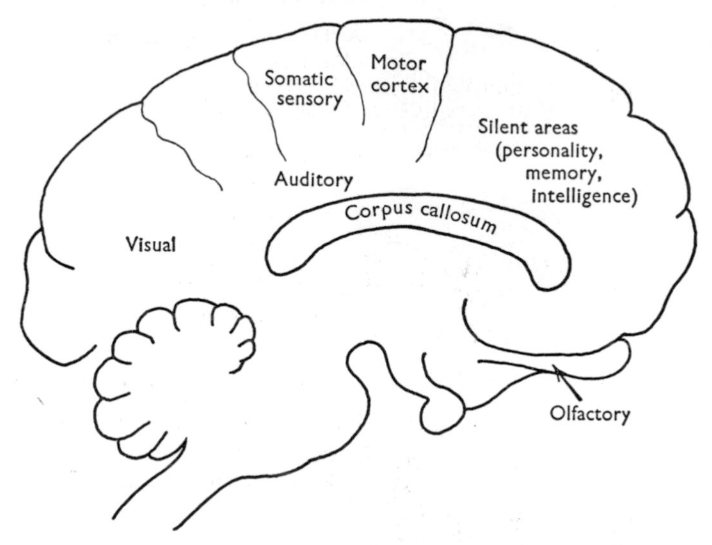

Fig. 132. Localisation of function in the cerebral cortex of man (shown in longitudinal section).

enormously enlarged and in man it almost covers the rest of the brain. It is very much folded on its surface, forming the gyri separated by grooves or sulci. The degree of folding seems to increase in the higher mammals, reaching its most complex form in man. The cortex contains an enormous number of cells (10^{10} have been estimated). As indicated above, it is a layered structure with at least six distinct layers being clearly recognisable. Most of these cells are relatively small, although in the motor area are found the largest cells—the Betz cells, which give rise to axons passing down the spinal cord as the *pyramidal tract*, which plays an important part in the control of voluntary movements. In addition to the many dendritic connexions of the cortical cells, there are also many axons passing to and from the surface. Some idea of their density is known from studies on the visual cortex of a cat. The number of afferent fibres was 25,000/mm^2 and the efferent fibres were three times as many. Some cortical cells have dendritic endings (apical dendrites) which reach to the surface and extend several millimetres along it. The functioning of this enormously complex structure is only scantily understood and presents one of the greatest challenges of modern biology. Considerable evidence indicates some localisation of function within the cortex (Fig. 132) and, as indicated

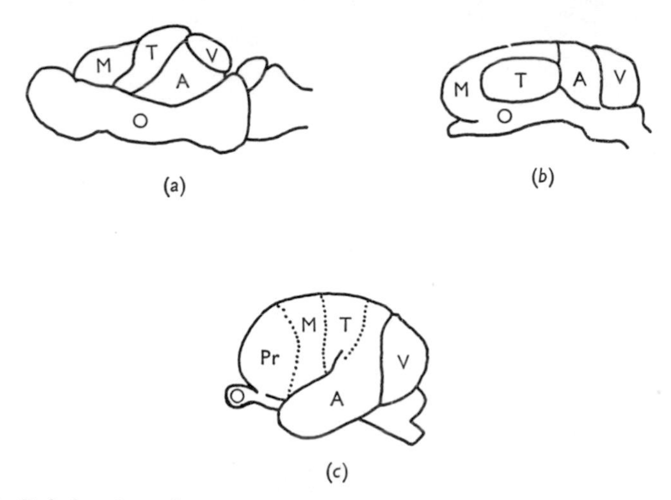

Fig. 133. Relative sizes of sensory projection areas. (*a*) Shrew; (*b*) pig; (*c*) monkey. Pr: prefrontal; M: motor; T: tactile; A: auditory; V: visual; O: olfactory.

above, one area (the motor area) is intimately concerned with movements of different parts of the body. For example, stimulation of localised regions in this area may produce very discrete muscular movements and all parts of the body seem to be represented in it. Just posterior to this voluntary motor area is a somatic sensory region, where electrical activity may be recorded when sensory stimulation is applied to different parts of the body. The representation in this area is not directly proportional to the size of the area concerned, but regions such as the hands and the face, notably around the mouth and lips, are relatively much greater in the size of cortex where responses may be evoked. This correlates with the observations derived from two-point stimulation mentioned earlier (p. 189). The visual area of the cortex is the projection area for visual pathways after they have passed through the thalami, others are concerned with auditory functioning and there are specific olfactory cortical projections. The relative sizes of these different sensory projection areas (Fig. 133) are closely related to the importance of the particular sense in the life of the animal. For example, in a pig the region of the cortex concerned with the tactile sensations of the snout is extremely large; the auditory area is very well represented in the dolphin. In different types of shrews the relative size of the visual area is correlated with their nocturnal or diurnal habits.

9 THE ENDOCRINE SYSTEM

9.1 The nature of endocrines

An endocrine gland is one specialised to produce a chemical substance called a hormone (derived from a Greek word meaning 'I arouse to activity') which passes to some near or distant area in the bloodstream. The affected region has its cells organised to respond in some way to the hormone concerned in their activation.

Hormone action is particularly concerned with metabolic activities—that is, anabolic or building-up processes and catabolic or breaking-down processes—in the cytoplasm. By altering the balance of these, hormones are able to coordinate such long-term changes as growth and maturation which are less suited to nervous control.

Both nervous and endocrine mechanisms use basic properties of the living cell, such as secretion and the propagation of impulses. It is not possible to decide which type of coordination came first in evolution and both methods are closely connected. Thus there are in arthropods and in the hypothalamus of the vertebrates certain neuro-secretory cells specialised for the production of hormones. In many cases neurones secrete a hormone for synaptic transmission. (Although by definition a hormone is a blood-borne substance and it is best to use the term 'chemical transmitter'.[1])

We can, in fact, distinguish three ways in which the secretion of a cell can be effective elsewhere. In the first case there are cells, such as the hypophysial tract from the hypothalamus to the posterior pituitary, which produce substances in the cell body which then pass out to distal processes of the cell where they are released. The second type are those, such as the neurones, which secrete at the distal endings only. Finally there are the endocrine cells which release substances in their immediate vicinity and depend on their transport in the blood.

[1] Many substances produced within the cell can affect the rates or nature of the reactions taking place in its protoplasm, for example, messenger RNA, and it can be seen how certain cells might have developed secretions affecting others remote from themselves.

So far as it is possible to generalise, the information passed by the hormones is necessarily a more gradual affair than that of the nerve impulse. There are, however, a number of instances of rapid endocrine coordination such as the reaction of appropriate areas to *adrenaline* or of the stomach to *gastrin*. Sometimes hormones may act antagonistically to nerve impulses and both systems make use of feedback mechanisms (see p. 2). The introduction of *thyroxine* or *insulin* into the body suppresses their release from the thyroid and pancreas respectively, while the build-up

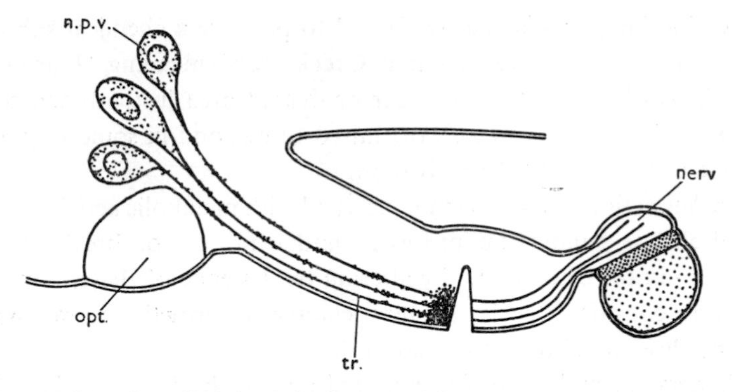

Fig. 134. Effect of sectioning of the hypophysial tract. Granules accumulate at the central ends of the cut fibres. nerv.: pars nervosa; n.p.v.: cells of paraventricular nucleus; tr.: hypophysial tract: opt.: optic nerve.

of *gonadal hormones* in the female during the maturation of the ovum suppresses the secretion of pituitary *gonadotropic* hormone.

During the lifetime of the animal the balance of hormones, sometimes called the hormone spectrum, is continuously changing and controlling the direction of its activities. Certain prolonged conditions such as exposure to stress or cold may lead to increase in the size of the endocrine concerned, in this case the adrenals, which is an adaptative change analogous to learning.

The mediator between the environment and the mammal, that is, the nervous system and especially the hypothalamus, is in close proximity to the *pituitary*. This is described as the 'master gland', and situated as it is below the fore brain it can be stimulated directly by external environmental change. The working of the pituitary is also related to the nearby hypothalamus which integrates so many of the visceral activities.

9.2 Outline of main endocrine organs and their secretions in the mammal

An outline will be found on pp. 232–4 (Table 13), in tabular form for convenience, while the way in which specific hormones act together in particular functions is dealt with subsequently. As representative activities, coordination of growth, metabolism, stress and reproduction have been taken.

9.3 Role of hormones in the coordination of growth and metabolism

The growth of a mammal represents a balance between protoplasm manufacture and its destruction; that is, between anabolic and catabolic processes within its cells. During early life the former are predominant and

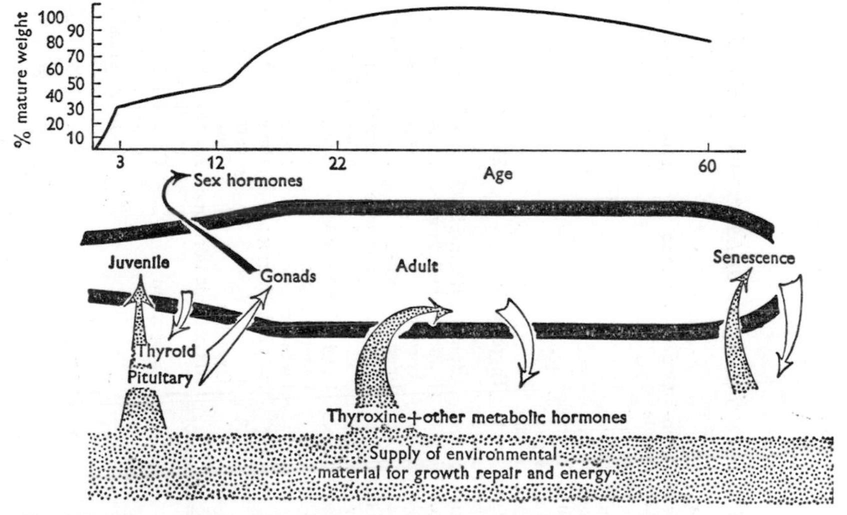

Fig. 135. Hormones and the life-cycle of man. Arrows from and to the environment indicate the extent of anabolic and catabolic processes.

material is rapidly incorporated into the body from the environment. The major influence during this period is the growth hormone from the anterior lobe of the pituitary which seems to encourage the removal and condensation of amino acids from the body fluids. These individual amino acids become condensed into the proteins of the living protoplasm. At the same time the metabolism of fats is also regulated so that the incorporation of fat and its laying down in the connective tissues is kept at a minimum.

Table 13. *A chart of the endocrine organs of the mammal with their origins, secretions and action on the body*

Organ	Origin and structure	Secretion	Effects	Excess and deficiency
Pituitary	The pituitary is situated below the hypothalamus of the fore brain and is made up of two parts. The *hypophysis* (anterior lobe) is derived from the roof of the buccal cavity, while the *infundibulum* from a down growth of the fore brain makes up the posterior lobe. Blood vessels pass from the hypothalamic region of the brain and form a plexus in the anterior lobe while the posterior lobe receives nerve tracts from hypothalamic and supraoptic parts of the brain. Being so near the teloreceptors and brain the pituitary can sample internal and external environment before the deeper endocrines. It is probably for this reason that it has come to play the role of master endocrine gland	Growth hormone	Stimulates growth by encouraging protein anabolism	If produced in excess during early life leads to *gigantism* of whole body, if later to abnormal development of hands, feet, jaw, etc. (known as *acromegaly*). If there is under-secretion *dwarfism* results as well as other symptoms associated with lack of thyroid and adrenal hormone
		Thyrotropic	Stimulates thyroid secretion	Results as for disturbance of thyroid (see below)
		Adrenocorticotropic	Stimulates adrenal cortex	Results as for disturbance of normal functions produced by secretions of these glands
		Follicle-stimulating hormone	Stimulates the development of ova and spermatozoa in the gonads	
		Interstitial cell-stimulating hormone	Stimulates the development of the interstitial cells of the gonads and the sex hormones they secrete	
		Luteinising hormone	Stimulates development of corpus luteum	
		Pancreotropic	Stimulates the growth of islet cells	
		Prolactin	Stimulates the secretion of milk	
		Metabolic hormones	Have direct effects on the metabolism of carbohydrates and fats	
	Posterior lobe	Oxytocin	Causes contractions of the uterus	
		Antidiuretic hormone	Increases the assimilation of water from the distal end of the kidney tubules	A deficiency of the hormone causes an overproduction of urine and consequent dehydration of the body. The condition is known as *diabetes insipidus*
		Vasopressin	Causes the constriction of the smooth muscles of the vascular system and thus increases blood pressure	

Thyroid	Originated in primitive chordates from the *endostyle* feeding groove in the floor of the pharynx which also had the function of iodine uptake and regulation. In mammals it consists of two lobes situated about the larynx, the lobes being composed of many alveoli with a secretory endothelium producing a fluid called the colloid which contains the thyroid hormone	Thyroxine	Stimulates the enzymes involved in respiration throughout the body; thus affects the whole metabolic rate, as well as the maturation of the mammal	In excess thyroxine produces a condition called *Graves's disease* with ex-opthalmic goitre and increase in the basal metabolic rate. This can lead to cardiac failure if prolonged. If congenitally deficient the lack of thyroxine causes *cretinism* where the individual fails to develop normally and the mind is retarded. There is also failure to develop sexually. Later in life deficiency, perhaps due to iodine shortage in the diet, produces a swelling of the neck or goitre and may lead to a slowing down of the metabolism and laying down of excess fat. The condition is known as *myxoedema*
Parathyroid	Originates, like the thyroids, from pouches that bud off from the embryonic pharynx. In man they are found embedded in the posterior part of the lateral lobes of the thyroid	Parathormone	The hormone causes a release of PO_4 from the kidney tubules which leads to its release from the bones and teeth. As the PO_4 ions are balanced by a simultaneous release of Ca the level of this ion is increased in the blood	Under-activity of the parathyroids causes a drop in blood Ca which in turn leads to muscular tetany. The over-activity of these glands would lead to a progressive demineralisation of the bones similar to rickets
Pancreas	In the mammal the pancreas is found as rather diffuse glandular tissue fitting in the first loop of the duodenum. However, in primitive chordates, e.g. lamprey, the pancreatic rudiments are found as cells in the wall of the gut and the regulation of sugar assimilation at this point probably evolved to its control throughout the animal. The pancreas has important digestive functions in mammals but the endocrine cells are clearly demarcated from the digestive and are grouped together as the islets of Langerhans (some 1–3 % of organ)	Insulin	Causes the utilisation and uptake of glucose by the tissues, thus reducing its level in the blood. (Antagonistic effects to adrenaline and glucocorticoids.) Insulin probably controls the tissue's permeability to glucose. Thus without insulin the glucose cannot be converted to fat, utilised in amino acid synthesis, or give up its energy	Failure to produce insulin leads to a condition called *diabetes mellitus*. The symptoms of this are high level of blood sugar, sugar in the urine, and a disturbance of the body's osmotic equilibrium and derangement of the nervous system. Toxic metabolites from fat (which need 'glucose energy' for their oxidation) also accumulate and are only lost from the kidney with valuable metal cations. The body becomes dehydrated. If excess of insulin is produced the utilisation of sugar is too great and its level falls in the blood which upsets nerve and muscle functioning

Table 13 (*cont.*)

Organ	Origin and structure	Secretion	Effects	Excess and deficiency
Adrenals	These consist of two different types of endocrine organ which in fishes (e.g. dog-fish) are found separate as long chains of secretory cells along the aorta. In the mammal the outer layer, or cortex, originates from mesodermal cells near the embryonic gonads, while the inner layer, or medulla, comes from cells associated with the formation of the sympathetic ganglia. The inner layer is characterised by taking up the stain chromic acid and for this reason its cells are called *chromaffin* cells	Cortical hormones	There are two main classes of cortical hormone, the glucocorticoids and the mineralocorticoids, which respectively conserve the sugar and mineral levels in the body. These hormones are also taken up by the tissues in stress conditions and help the body to withstand such conditions. Sex hormones are also made in small quantities by the adrenal cortex	The destruction of the adrenal cortex, such as occurs in *Addison's disease*, will lead to general metabolic disturbance, in particular weakness of muscle action and loss of salts. Stress situations, such as cold, which would normally be overcome, lead to collapse and death. The reverse of this is found in *Cushing's disease* where too much cortical hormone is produced. Symptoms are an excessive protein breakdown into glucose caused by glucocorticoids, and resulting muscular and bone weakness. The high blood sugar disturbs the metabolism as in diabetes
		Adrenaline	This hormone, and the associated noradrenaline, are released into the blood in stress conditions. Their effects are to stimulate the sympathetic system and bring the body into a state of readiness to respond and perform efficiently in 'fight or flight' reactions to the stress situation	
Gonads Ovary	The ovaries consist of a germinal epithelium and a large number of follicles in various stages of development. The various hormones associated with the ovary are as follows:			
	From the developing Graafian follicle	Oestrogen	Causes development of the secondary sexual characters in the young mammal, in the oestrous cycle leads to enlargement of uterus and its lining as well as the mammary glands. Inhibits FSH and stimulates LH. Brings about changes associated with pregnancy such as enlargement of the uterus and mammary glands, as well as metabolic changes. It also inhibits FSH	
	From the ruptured follicle (corpus luteum)	Progesterone		

Once the juvenile period is passed, the *adult* phase establishes itself and the hormone spectrum maintains the body in a state of metabolic equilibrium with anabolic and catabolic processes approximately balanced. Meanwhile the influence of the gonadal or sex hormones causes changes in the anatomy and behaviour of the individual.

The last phase is that of *senescence* and once again this is influenced by hormonal change. In this phase, degradation of protoplasmic substances overtakes their anabolism and the rate of metabolic change and replacement is reduced (Fig. 135).

At each of these stages of life the hormones interact together to produce their effects. Thus during early development the activity of the pituitary is only made possible by the presence of thyroxine and *corticosterones*, which together prepare the body cells to respond to the master growth substance. Throughout the life of the individual the sex hormones and those of the pancreatic islets also encourage anabolic metabolism.

Not only do the endocrines bring about gross changes in the metabolism of mammals over long periods but they also regulate at any one time the rate of physiological processes taking place. A good example is insulin, from the islet cells of the pancreas, which encourages the uptake and condensation of sugars from the blood by the tissues, and by doing this it restricts the amount of sugar available for respiration. Insulin promotes the synthesis of fats within the tissues of the body from the fats and fatty acids in the blood and lymph. (It should be noted that the pancreas produces another hormone called *glucagon* which increases the blood sugar level but does not directly antagonise insulin. Hence a disturbance of pancreatic functioning, as in *diabetes,* may produce very complex symptoms.)

Thyroxine is another hormone which causes rapid changes in the metabolic rate of the body, its function being to facilitate the normal operation of a number of respiratory enzymes and thus control energy release. At the same time corticosterones regulate the build-up and break-down of amino acids and their final deamination and respiration.

Besides the major endocrine roles outlined above a number of subsidiary activities will be taking place in the endocrine system. In the first place certain of the secretions of the anterior lobe of the pituitary, in the form of the tropic hormones, stimulate all the above endocrine organs, so exerting an overall control. It is also known that the anterior pituitary affects metabolism directly, and hormones it produces, such as ACTH (*adrenocorticotropic hormone*), cause the release of fatty acids from tissues into the bloodstream. Similarly the pituitary *ketogenic* hormone controls the level of fat respiration.

Other, and more recently described hormones, affect blood pressure, such as *angiotensin* from the kidney and *antidiuretic hormone* from the pituitary. A further substance called *serotonin*, originally isolated from wound tissues, causes contractions of blood vessels, while *histamine*, produced in allergic responses, increases capillary permeability.

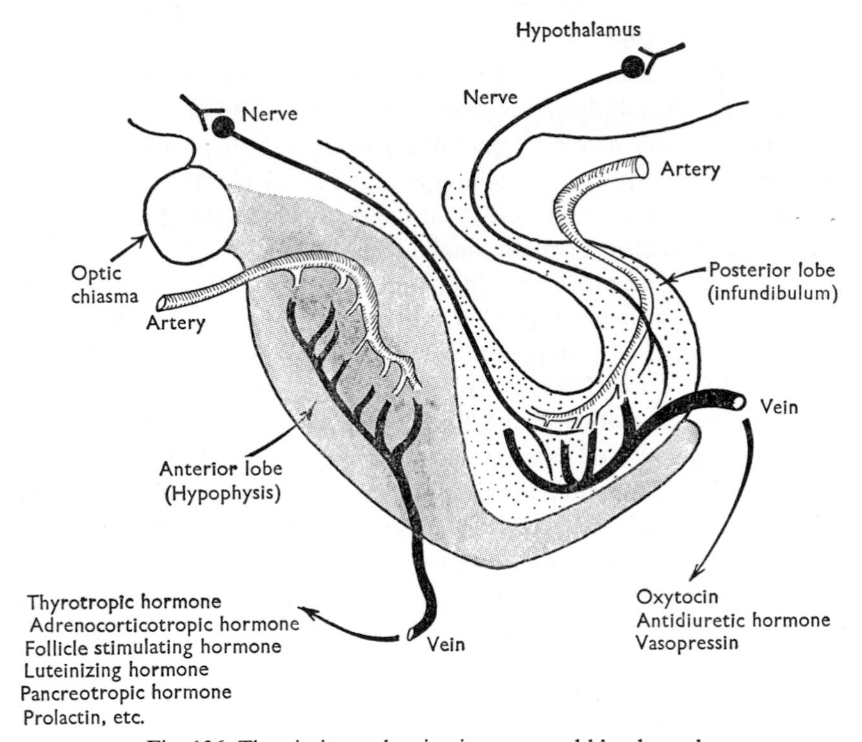

Fig. 136. The pituitary, showing its nerve and blood supply.

9.4 Hormones and stress reactions in the body

A particular type of situation which is largely coordinated by hormones is the body's reaction to stress, which may be in the form of pain, fear, heat, cold, anger, etc.

When the individual is confronted with a situation of this sort impulses from the sense organs pass to the cerebral region and thence to the hypothalamus. From here they are relayed to the pituitary, which produces ACTH, and to sympathetic neurones in the spinal cord. Motor impulses

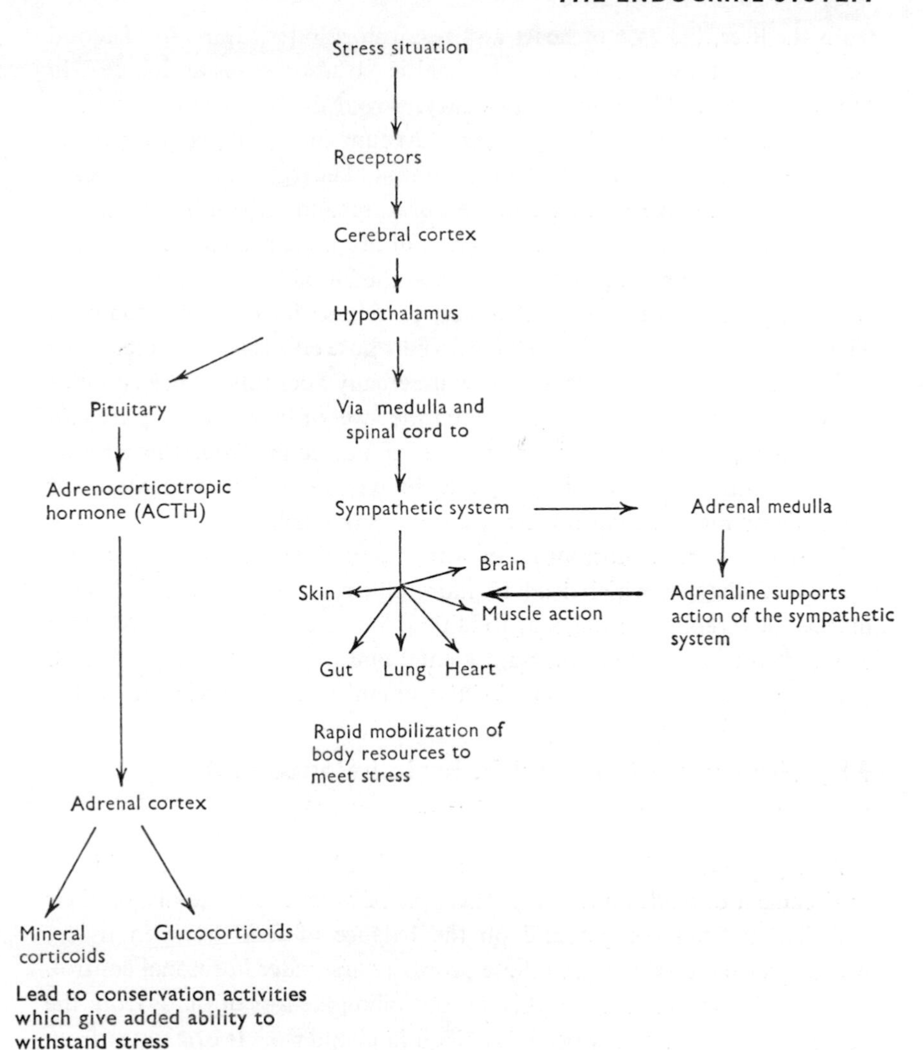

Fig. 137. Sequence of events in the body's reactions to stress situations such as danger or cold. Activation of the adrenal medulla precedes that of the cortex.

travel out to the sympathetic system and also to the adrenal medulla which secretes the hormone adrenaline.

Adrenaline reinforces the action of the sympathetic system and prepares the body to meet stress. Because of its effects it is known as the 'fight or flight' hormone, preparing the body for one of these responses according to the nature of the animal. Thus adrenaline produces the release of sugar from the liver, increase of heart and respiratory rates, decrease of blood flow to the gut and inhibition of peristalsis. It also causes an increase in the release of ACTH from the pituitary. From the biochemical point of view the action of adrenaline promotes the functioning of *phosphorylating enzymes*, leading to an increase in the amount of energy available. (Associated with adrenaline is the hormone nor-adrenaline which has much the same effects on the body as the former but at different concentrations.)

The ACTH from the pituitary passes in the blood to the adrenal bodies where it stimulates the cortical region to release its hormone complex. The best known of these is a glucocorticoid whose effects are to increase the resistance of the body to stress. How it actually does this is very complex but it is known that wound healing and the level of blood glucose are two of the things affected. A further class of hormones from the adrenal cortex are the mineralocorticoids, and these conserve Na^+ and water for the body by their action on the renal tubules (see below).

Thus the role of hormones in stress situations is to bring about immediate changes which lead to more efficient functioning of skeletal muscle and its coordination, and to initiate long-term changes whereby the effects of the stress situation, both mental and physical, can be repaired. (This is an acclimation response of the organism to its environment.)

9.5 Hormones involved in ionic balance and osmo-regulation

At any one time the regulation of a constant internal osmotic and ionic environment depends on a sufficient supply of water and minerals, on their assimilation from the gut and on the balance of fluid between tissue, blood and excretory loss. All these processes are under hormonal control.

The antidiuretic hormone (ADH) controlling the loss of fluids from the kidney tubule has already been described in chapter 6. It originates from the posterior lobe of the pituitary and determines the permeability to water of the cells of the distal end of the tubule.

Besides this ADH the water balance is maintained by the adrenal-corticoid hormone *aldosterone* which inhibits water uptake from the tissues as well as preventing its elimination from the kidney. This hormone also increases the uptake of Na^+ from the gut and inhibits its removal from the kidney so it conserves this important ion. Vitamin D, which, if made in the body, may be considered a hormone, causes the assimilation of Ca^{++} and PO_4''' from the gut and its release from bone. These ions are also regulated by *parathormone* from the *parathyroids*, small endocrine glands near the thyroid, which when stimulated by lack of Ca^{++} in the blood cause increase of its absorption from the gut, as well as its retention in the kidneys. The parathormone also causes release of both Ca^{++} and PO_4''' from bony tissues but as it encourages the elimination of the latter ion from the tubule it actually results in a rise in Ca^{++} ions only.

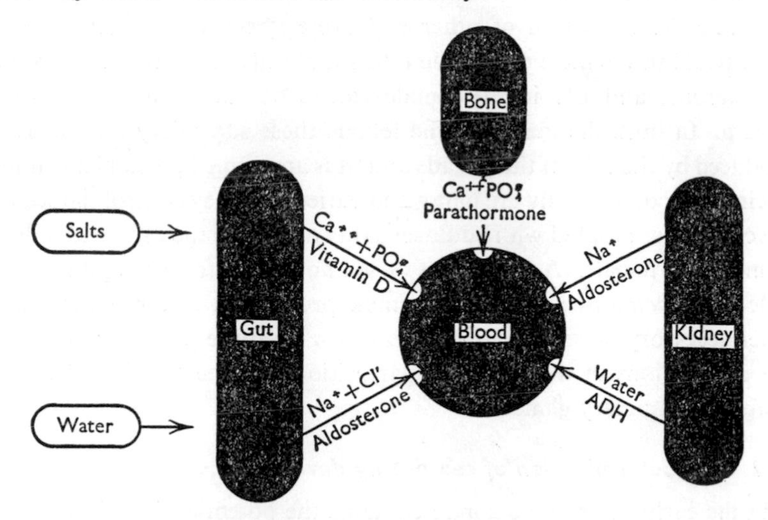

Fig. 138. Diagram summarising the action of hormones in osmotic and ionic regulation.

9.6 The role of hormones in reproduction

9.61 *General activity of sex hormones*

The young mammal may be thought of as concentrating its activities towards growth and the acquisition of behaviour patterns which will enable it to survive away from the parental care it enjoys after birth. In social animals and those with few young, such as man, the period may be prolonged, but, sooner or later, the adult state is reached and an independent life gained.

During this transition period between the juvenile and adult the hormone balance which has encouraged rapid growth and the retention of juvenile characteristics also changes, and endocrines from the anterior pituitary begin to act on the gonads. This in turn leads to further secretion of hormones which bring about the drives and secondary sexual characters associated with the adult.

In the first part of this maturation sequence hormones from the pituitary called *gonadotropins* are responsible for the changes taking place in the gonads. It is convenient to divide the gonadotropins into those, such as *follicle-stimulating hormone* (FSH), which act directly on the gonad and stimulate development of the gametes, and those, such as the *luteinising hormone* (LH) or *interstitial cell-stimulating hormone* (ICSH), which stimulate the formation of other endocrine systems within the gonads. Thus ICSH in the male causes the interstitial cells of the testes to produce testosterone, and LH in the female causes development of the corpus luteum. In both the male and the female these subsidiary hormones are produced by the cells of the gonads and thus are termed gonadal hormones. Their effects on the body are many and varied and they control the secondary changes associated with adult sexuality as well as the differential drives of male and female. A further set of sex hormones from the pituitary are called the *lactogens* and stimulate milk production in the female, while other accessory hormones such as *oxytocin* from the posterior lobe cause the contractions of the uterus at parturition and the 'let down' of milk from the mammary gland.

9.62 *The determination of sex during development*

In the early embryo the gonad contains the potentialities of both sexes. The outer layer appears to correspond to the female part and the inner area to the male part. The sex chromosomes (i.e. the genotype) that the individual inherits will either cause the development of male or female sexuality. Once activated the sex cells of the one type produce hormones that bring about development along the appropriate morphological lines and inhibit the development of the other sex.

This means that the sex of an individual is a response to a particular hormone balance, normally determined early in embryological development, but also subject to variation during adult life. Extreme changes in later life, such as sex reversal, are rare at least in mammals, but it should be appreciated that the difference between maleness and femaleness is one of degree only. Abnormal gene complexes, such as an extra X-chromosome in a male, or hormonal disturbances, can lead to a condition of intersex.

9.63 *Sex hormones in the male*

As described above, the pituitary hormones have two effects on the male. In the first place the follicle-stimulating hormone (FSH) causes the spermatozoa to develop from the lining cells of the *seminiferous tubules*, and in the second place luteinising hormone (LH) or ICSH stimulates the release of *testosterone* from the interstitial cells of the testes.

Despite this fairly simple picture of the role of sex hormones in the male it must be appreciated that there are a number of subsidiary hormones and effects. Thus the *androgamones* of the prostate are important in determining the activity of the sperm, the adrenal cortex produces male hormones, and side effects of testosterones produce parental behaviour in the male.

9.64 *Sex hormones in the female*

The sex hormones of the female are more complex than those of the male, as more activities have to be coordinated. Beside the follicle-stimulating and luteinising hormones from the pituitary, which promote the development of the ovary and secondary sexual characters respectively, a number of other hormones coordinate physiological changes during pregnancy, birth and post-natal care of the young.

The control of the egg-producing mechanism, or *oestrous cycle*, in the female is a good example of a rhythmical event brought about by endocrine interaction. (In the human female this cycle is not initiated by changes in the environment although it may be affected by such changes.)

The oestrous cycle is initiated by the release of follicle-stimulating hormone (FSH) from the pituitary which causes the development of the next follicle due to ripen. The ripening follicle procudes the gonadal hormone *oestrogen*. This encourages follicle development but inhibits the production of the FSH, which ceases. At the same time the oestrogen causes the pituitary to release luteinising hormone (LH) which, passing to the ovary in the bloodstream, produces ovulation and the change of the ruptured *follicle* into a further endocrine organ. This new endocrine is the *corpus luteum* and forms by the cells of the burst follicle becoming secretory to produce the female hormone *progesterone*.

Progesterone has two effects: in the first place it further inhibits FSH (another example of negative feedback) thus preventing follicle development, and in the second place it prepares the female for the retention and nutrition of the foetus. If the ovum is fertilised and becomes implanted on the *uterus* wall, the corpus luteum goes on producing progesterone until its role is taken over by the *placenta*. On the other hand if the ovum is not fertilised and thus neither oestrogen nor progesterone levels are maintained once again FSH commences to flow from the pituitary and the whole cycle recommences (see Fig. 140).

At parturition the supply of progesterone from the placenta declines,

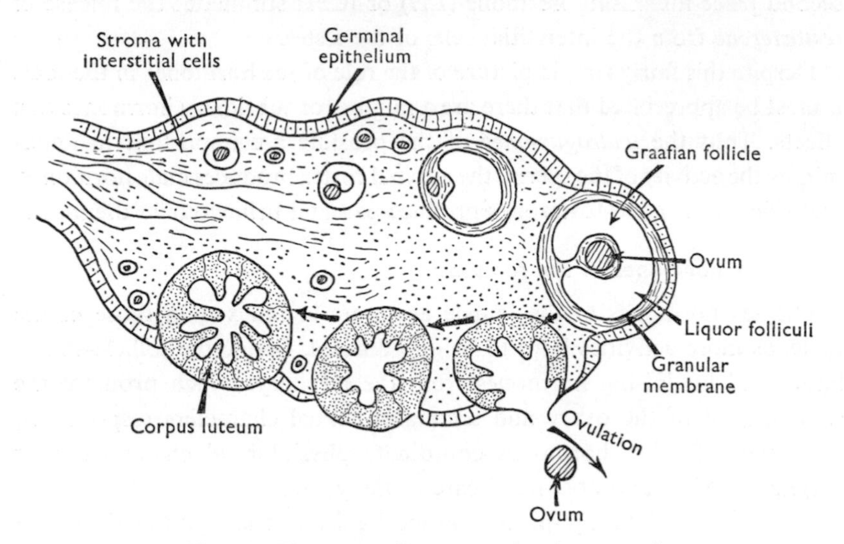

Fig. 139. The mammalian ovary, showing stages in development of follicle.

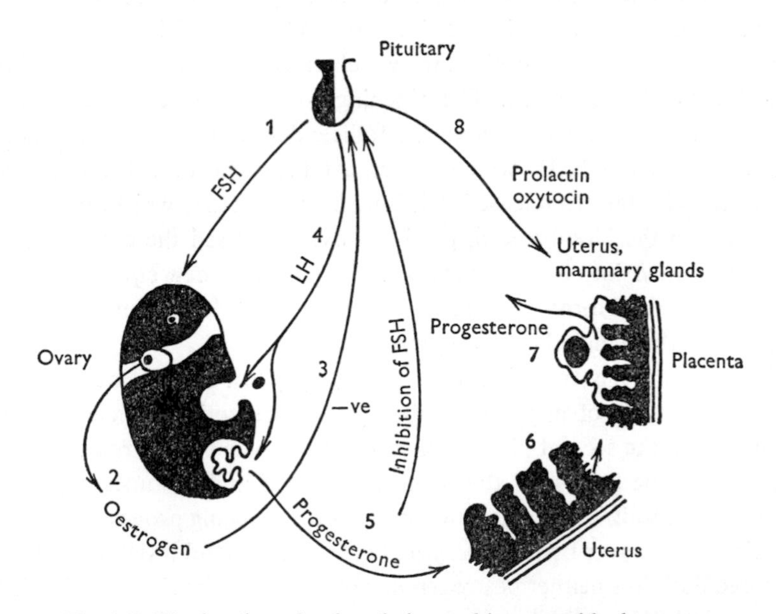

Fig. 140. The female cycle of ovulation and its control by hormones.

allowing oestrogen and posterior pituitary hormone, which cause contractions of the uterus, to take effect. After birth, changes in the progesterone concentration, together with a shift in the amounts of other hormones, for example thyroxine and the adrenal complex, and the stimulus of the suckling infant, cause the release of prolactin from the anterior pituitary. The latter causes the secretion of milk from the mammary glands, whose cells have been enlarging during gestation. The prolactin also contributes to the drive of the female towards the care of her offspring.

9.7 Examples of hormone action in vertebrates other than the mammals

As already shown above, the endocrine organs of mammals can, in most cases, be homologised with those of other vertebrates, and the main endocrine controls described operate throughout the whole of the Craniata. On the other hand there are some examples of hormonal coordination seen in the lower vertebrates which are not commonly met with in mammals. Among these are the control of colour change, migration and metamorphosis, where they occur.

9.71 *Fishes*

9.711 *Metamorphosis.* Some fishes such as the eel and plaice have an immature form which is very different from the adult. The change from such a larval stage into the adult is called metamorphosis and is under control of thyroid hormone.

9.712 *Salinity.* Salinity tolerance in fishes such as the eel which move from sea- to freshwater or vice versa is under control of the posterior pituitary and adrenal cortex. The ability of arctic cod to enter cold water has recently been shown to be governed by the thyroid.

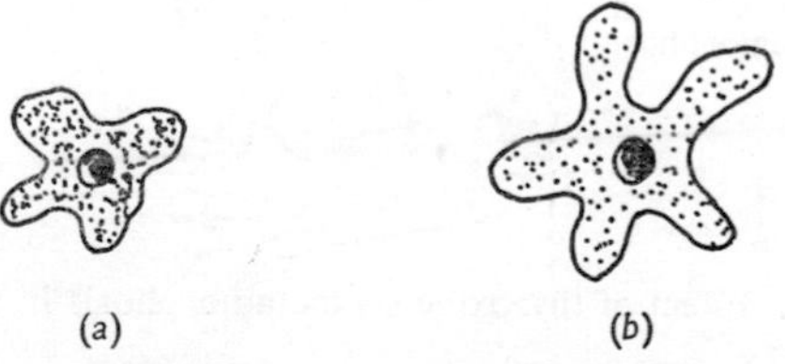

Fig. 141. Melanophore. (*a*) Contracted; (*b*) expanded.

9.713 *Colour changes.* In many species of fishes, colour changes are due to the secretion of a melanophore-dispersing or melanophore-condensing hormone from the posterior pituitary. The stimulus for the release of these substances is via the eyes and brain to the pituitary. The two hormones act on the melanophores, amoeboid cells in the skin, and, by the spreading out or contraction of the brown pigment, produce camouflage patterns which blend the fish into its natural background. In many Teleosts colour change is under nervous control. Some demersal species such as the turbot can make a remarkably accurate fit to complex external patterns (e.g. a chess board).

9.714 *Changes associated with sexual dimorphism.* The androgenic hormones of the male (that is, sex hormones from the gonads) are responsible for bringing about the changes in colour or other features seen at the breeding season. A well-known example is the bright colour of the male stickleback which develops during the reproductive period.

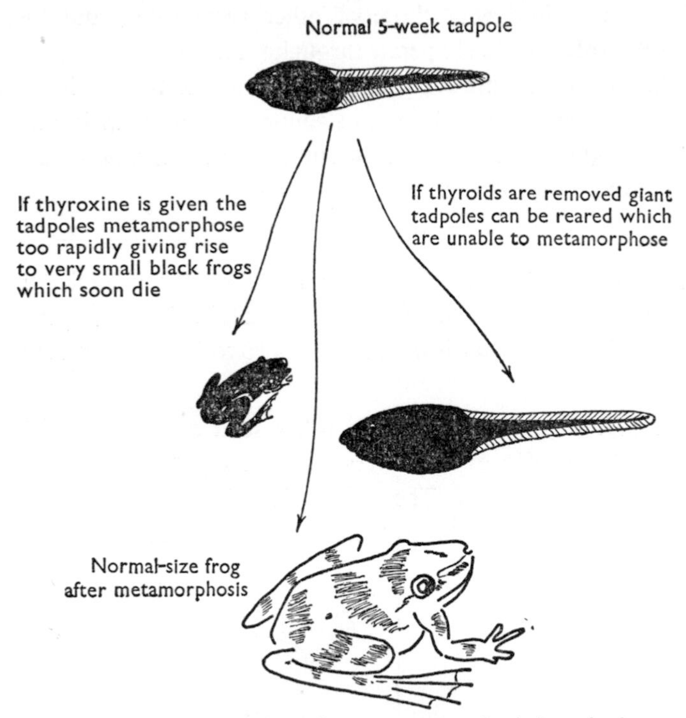
Normal 5-week tadpole
If thyroxine is given the tadpoles metamorphose too rapidly giving rise to very small black frogs which soon die
If thyroids are removed giant tadpoles can be reared which are unable to metamorphose
Normal-size frog after metamorphosis

Fig. 142. Effect of thyroxine on metamorphosis in tadpoles.

9.72 Amphibians

9.721 *Metamorphosis.* The change of the tadpole into the adult frog or toad is under the control of the hormone thyroxin and for this to be secreted it is necessary that thyrotropic hormone is released from the anterior lobe of the pituitary. By depriving the developing tadpole of iodine, necessary for the synthesis of thyroxine, or by removing its thyroid gland, metamorphosis can be prevented. By treatment of young tadpoles with thyroxine precocious metamorphosis is brought about.

9.722 *Colour changes.* In the amphibians colour changes are also based on the contraction or expansion of pigment within the melanophores of the skin and as with fishes these movements are controlled by antagonistic pituitary hormones.

9.73 *Reptiles*

9.731 *Colour changes.* In many lizards, such as the iguana, pituitary and adrenal hormones, as well as nervous mechanisms, control colour changes. For others, such as the chameleon, the control is exclusively nervous.

9.74 *Birds*

9.741 *Reproduction.* The differential behaviour of the male and female during the reproductive period has been associated with gonadal hormones. An interesting subsidiary hormonal effect is by the pituitary lactogen which causes the secretion of 'milk' from the pigeon's crop.

9.742 *Migration.* The removal of the gonads does not prevent migration taking place so that the drive to migrate must not be directly associated with ripening gonads. It is thought that the gonadotropic hormones from the anterior pituitary whose secretion is stimulated by the lengthening day also change the physiological state of the bird. One of the changes is the deposition of fat, which acts as the 'fuel' in migratory species (it is an economical substance to store), and it may be this latter change which acts as the direct stimulus to migration.

10 REPRODUCTION

10.1 Introduction

The life of an animal may be considered as a single chain from its conception until it has produced its kind. As far as biological survival is concerned the post-reproductive phase of life is unimportant for the majority of animals and, in nature, few survive long after their reproductive capacity is exhausted. In fact successful reproduction is the main 'object' of a species, so during evolution we might expect this physiological process to have come under the most rigorous selection.

For reproduction to be successfully carried out a number of events must occur. In the first place the balance of hormones must shift from encouragement of growth towards the stimulation of the gonads and assumption of the secondary sexual characters. There must develop both the means of reproduction—that is, gametes and sex organs—and also the drive to co-operate with another individual to ensure fertilisation. In the second place some sort of provision must be made regarding both the number and the nutrition of the *zygotes* produced. A satisfactory balance must be struck between the large number of offspring needed to survive the natural selection of their environment and the distribution of a limited food supply among a large number. Finally the efficiency of reproduction is much increased where there is utilisation of the parental maturity to protect the young before or after birth.

During evolution there has been a trend of increasing efficiency in the reproductive processes of vertebrates and parallel adaptations with lower animals are found. Thus the progressive provisioning of larvae that is seen in the *Hymenoptera* and other insects is paralleled by the *post-natal feeding* of the offspring in birds and mammals. Although the general tendency towards a reduction of the number of young and increasing parental care more or less follows the stages of vertebrate evolution there are many lower vertebrates that show convergent adaptations with birds and mammals (Table 14). Despite these occasional similarities there can be no doubt that the colonisation of the land was the major environmental change affecting the reproductive processes of vertebrates and that it was associated with a number of adaptations. Thus internal fertilisation, the *cleidoic* (enclosed in shell) egg, the foetal membranes, the control of temperature, which all affect reproduction, might never have been selected in the constant environment of the sea.

Table 14. *The number of eggs laid by different vertebrates*

Animal	Eggs produced at one time
Fish	
Cod	3,000,000–7,000,000
Herring	30,000
Amphibian	
Frog	1000–2000
Reptile	
Adder	10–14
Bird	
Pheasant	14
Thrush	4–5
Mammal	
Dog	4
Man	1

For comparative purposes we may consider in outline the typical methods of reproduction within each vertebrate class, drawing attention to the special adaptations that have been made at each level. The section relating to the mammal should be read in conjunction with pp. 239–242 of the last chapter which described the role of endocrines in reproduction.

10.2 Fishes

Most teleosts, such as plaice or trout, produce a very large number of yolky eggs at a time which is appropriate to the food supply in the environment. The stimulus for ripening of gonads is supplied by pituitary hormones which depend for their secretion on rhythmical environmental change. Nearly all fishes show *antenatant* (against the current) spawning migrations so that they actually release eggs up-current of the feeding grounds, which allows the larvae to drift down towards the latter. The majority of sea fish, an exception being the herring, have *pelagic* or floating eggs, while most river fish make nests or secure their eggs in some way to prevent them drifting downstream. To produce coordinated mating, elaborate shoaling rituals may take place (e.g. cod) or there may be patterns of behaviour between pairs (e.g. stickleback) ensuring the release of gametes at the same time.

Fishes, on the whole, show little parental care for their offspring and in fending for themselves a great mortality of young takes place. It has been shown for the North Sea plaice that the young fish must have a certain type of diatom food within a few days of hatching if it is to survive.

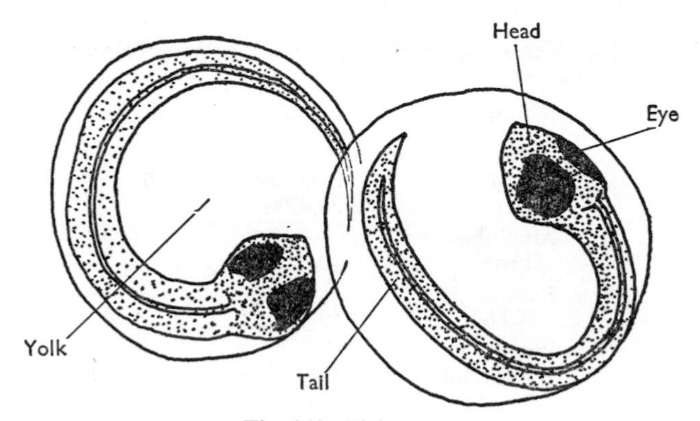

Fig. 143. Plaice eggs.

There are some exceptions to the above reproductive methods. The elaboration of nests occurs in certain river fish and, as in the case of the stickleback, this provision may extend to protection of the young after hatching. There are catfish and species of sea-horse which retain their young in their mouths or in brood pouches, while others, such as the guppy, are viviparous and the eggs develop within the body of the female.

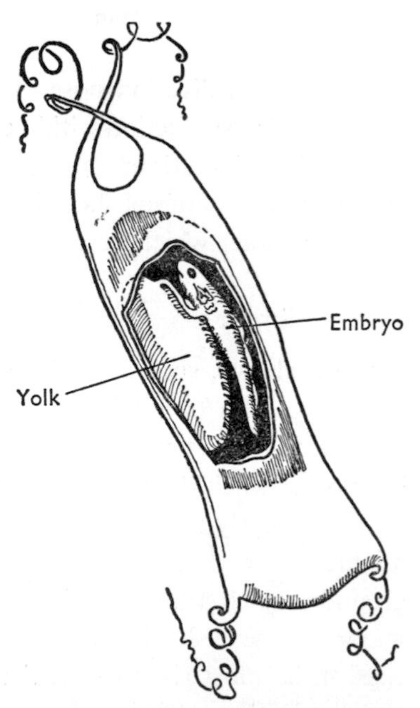

Fig. 144. Dogfish's egg (mermaid's purse), showing position of embryo.

This latter system is well developed among the elasmobranchs (e.g. *Squalus acanthias*) and it involves a means of internal fertilisation, the *claspers* of the male, and allows a small number of eggs with a quantity of yolk to be produced. In the common dogfish (*Scyliorhinus canicula*) internal fertilisation is followed by the secretion of a protective case around the eggs and this is attached by its trailing threads to objects on the sea bed.

In some elasmobranchs an ovoviviparous condition has evolved with some degree of assimilation of nutrient from the uterine-wall secretions of the mother. Such is the case in the smooth hound, *Mustelus vulgaris*, whose developing embryos take up mucoprotein, fat and monosaccharide sugar. The uterine secretions also contain urea, which forms an important part of the embryo's environment.

Again some elasmobranchs have true vivipary and a placenta is formed between the wall of the uterus and the yolk sac. In fishes such as *Mustelus laevis*, the weight of the embryo increases several hundred per cent during development. Where a placenta is formed, the wall of the uterus may have large numbers of villi, and projections from the yolk sac burrow into this to make a large surface bond between parent and embryo.

A transfer of substances between the female elasmobranch and its developing young is associated with a severe loss in weight of the maternal liver during gestation. This is true of both those that form a placenta and those that secrete nutritive fluids. Loss of liver weight does not take place in mammals during gestation.

10.3 Amphibians

Stimulated by increasing daylight and other factors the endocrines of the frog which control reproduction become active during the latter part of hibernation. Metabolic changes occur and the quantities of stored fat decrease while the gonads enlarge and ripen. Most frogs return to water for reproduction and they may migrate several miles to find their original spawning ground. The croaking of the male frog and the enlarged belly of the female are two of the stimuli which lead to mating, which is accomplished by the male grasping the female with his special nuptial pads. Fertilisation of the eggs takes place externally and the sperm must penetrate the egg rapidly before its coating of albumen swells. In the male the sperm pass down from the testes via the anterior part of the kidney and the Wolffian ducts. Both sexes have a *cloaca* where genital as well as excretory products are passed to the exterior. All these features as well as the need to return to water indicate the primitive nature of amphibian reproduction.

Once laid, the eggs have the limited protection of their albumen coating and the fact that the black pigment they contain has a bitter taste. The larval amphibians or *tadpoles* are well adapted to life in water, having respiratory and locomotory systems as well as sensory adaptations, such as the lateral line, which are similar to those of fishes.

While this lack of parental care is typical of the majority of amphibians there are a good number of exceptions. The main adverse feature of the reproductive system described above is its necessity for external water. Some tree frogs 'sew' or stick leaves together, making hollows high above the ground where rain water collects and into which they can lay eggs. Among the most remarkable adaptations, however, are those of the female midwife toad which has a number of holes in her back into which she pushes the eggs after fertilisation. In these pockets the tadpoles develop.

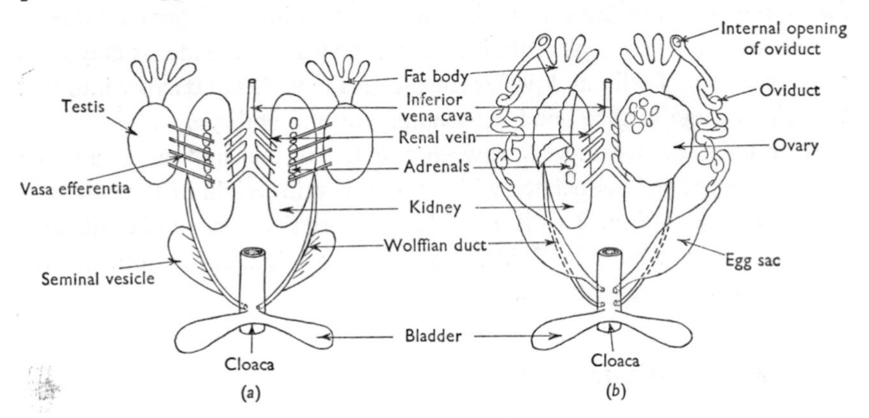

Fig. 145. Diagrams of urinogenital systems of frog. (*a*) Male; (*b*) female.

10.4 Reptiles

With the more efficient colonisation of the land achieved by the reptiles we see many changes from the amphibian condition. Fishes and amphibians are classed as *anamniotes* as they do not possess the *amnion* characteristic of the reproductive mechanism of reptiles, birds and mammals. This amnion is an *extra-embryonic membrane* evolved in conjunction with the shelled egg and, by its means, the embryo is able to develop within a stable fluid environment. In conjunction with the amnion a further extra-embryonic membrane, the *allantois*, develops; this has a role in respiratory exchanges and excretion. As with anamniotes there is yolk enclosed in a sac and this provides the raw material for the developing embryo. The albumen of the egg contains water and the porous shell allows gaseous exchange, so that at the level of the reptiles a great deal of provision is made for both the nutrition and protection of the developing offspring. *Parental care* given after birth is very limited, however, and the number of eggs that have to be laid to ensure a reasonable replacement rate is high.

Snakes guard their eggs and turtles and crocodiles, among others, bury their eggs to provide them with a uniform and protected environment.

A few lizards and some snakes have developed vivipary but there is no exchange of material between mother and offspring.

10.5 Birds

The egg of a bird is much the same as that of the reptiles from which the birds evolved, but there is a wider range of colour and shape. Birds with open nests usually have camouflaged eggs (e.g. plover) while those with concealed or domed nests have white eggs (e.g. owl and long-tailed tit). There is a general tendency to reduce the number of eggs laid in the more advanced birds, thus the ostrich has some 80 eggs while the highly successful fulmar has only one, but on the whole birds lay far fewer eggs than do reptiles. Recent work has shown that the number of eggs laid by a bird is related to the number of offspring that the parents can successfully provision. Variations exist within a given species according to the date of nesting, the latitude and immediate ecological conditions.

Parental care is well developed over the whole reproductive period and the male and female tend to cooperate in the making of the nest, incubation of the eggs and subsequent care and feeding of the young. Nest sites are chosen to give the maximum protection from predators and for insulation soft materials such as down and moss are used. The actual position of the nest in regard to other members of the species is also important. Land birds have quite extensive *territories* which they defend and which have the effect of dispersing a given species efficiently in an area. Sea-birds have their nests closely placed for protection against predators but still regard the small area around their nests as territory.

Birds being homoiothermic it is essential that they maintain their eggs at constant temperature. The construction of the nest assists in this and the female (and sometimes the male) develops an increased blood supply to the skin as well as losing feathers from her breast. She also develops the drive to incubate, which is very strong and can be clearly seen in the behaviour of a 'broody' hen.

After hatching the young are usually cared for by both parents, the gape of the fledgling's beak acting as the releaser to the parents' feeding responses. *Nidicolous* young are slow to develop and remain for a long time in the nest, which will tend to be built away from predators. *Nidifugous* young develop rapidly and soon leave the nest, which will be sited on the ground. Most passerines come into the first class while game birds such as duck and pheasant fall into the latter.

The interesting feeding method of the young pigeon is described on p. 245.

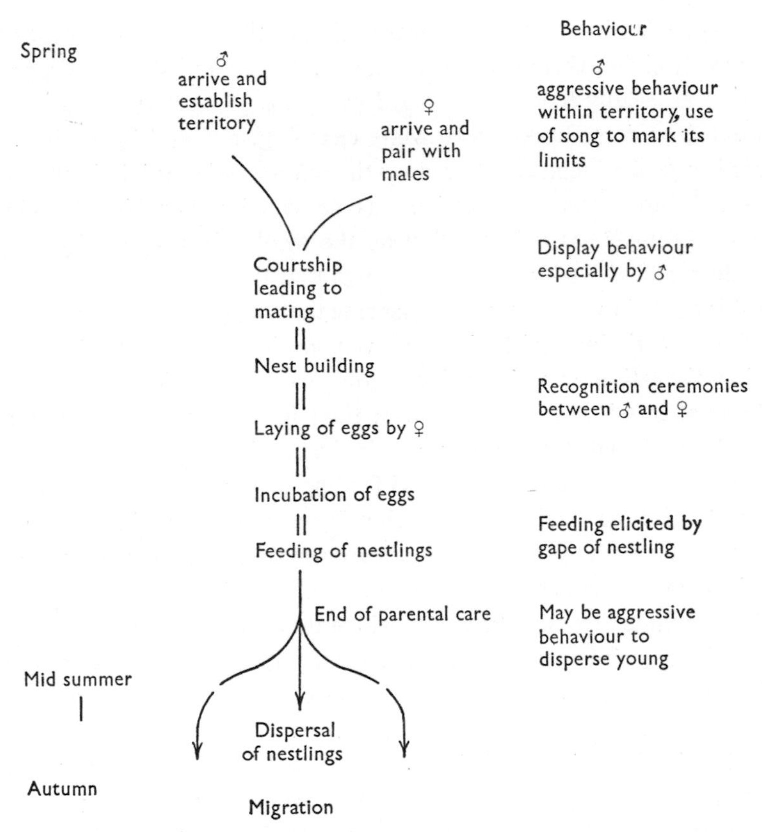

Fig. 146. Scheme to show how many birds cooperate over a period of months for reproductive purposes.

10.6 Mammals

Together with the development of their cerebral hemispheres and complex behaviour the mammals owe their success to the efficiency of their reproductive system. In this class of vertebrates we find protection of the developing young and subsequent care of the offspring exceedingly highly developed.

The egg, which still goes through a similar gastrulation to the reptile and bird egg, is quite small in the placental mammals although the shelled eggs of the primitive monotremes are large. Mammalian eggs have a reduced yolk because they are retained in the female body and nourished from her tissues. In order to do this the mother's oviduct has a specially modified and muscular region called the *uterus*, and between the developing embryo and the mother is formed the *placenta* for the exchange of metabolites.

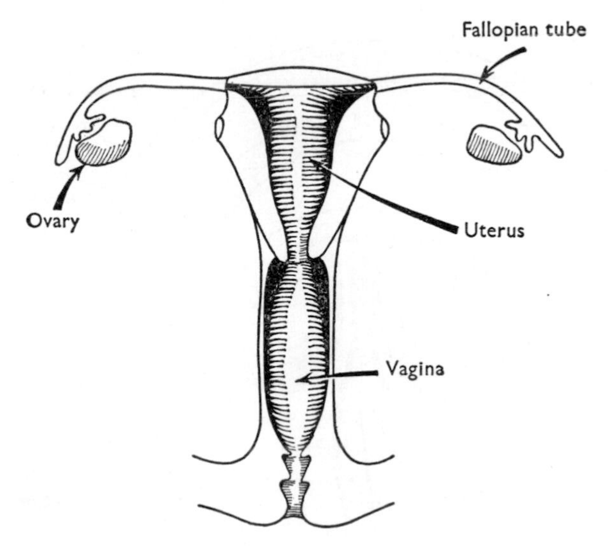

Fig. 147. Human female reproductive organs (opening of urethra not shown).

The placenta forms from the extra-embryonic membrane, the *chorion* (the outer of the amniotic folds which surround the embryo—see Fig. 148) coming into close contact with the lining of the uterus. Later the allantois grows out from the foetal endoderm and fusing with the chorion gives rise to the *chorio-allantoic* placenta. Both the inside of the chorion and the outside of the allantois are lined with mesoderm and in this blood vessels arise forming the *umbilical artery and vein.* The foetal heart drives blood up the umbilical artery and between foetal and maternal circulations a counter-current exchange system may develop (see Fig. 48); food and water and oxygen are passed from the maternal circulation and carbon dioxide and other metabolites returned. There has been a tendency to reduce the layers involved in the placenta in many orders of mammals, thus, in our own case, the maternal epithelium, connective tissue and endothelium are not present in the placenta and the foetal tissues project directly into the blood sinuses. In such placentas the uptake of ions and presumably other substances is greatly speeded up and in this respect they are more efficient than those in which some of the layers become lost during development. There is no reason to suppose that species with a many-layered placenta are necessarily the most primitive.

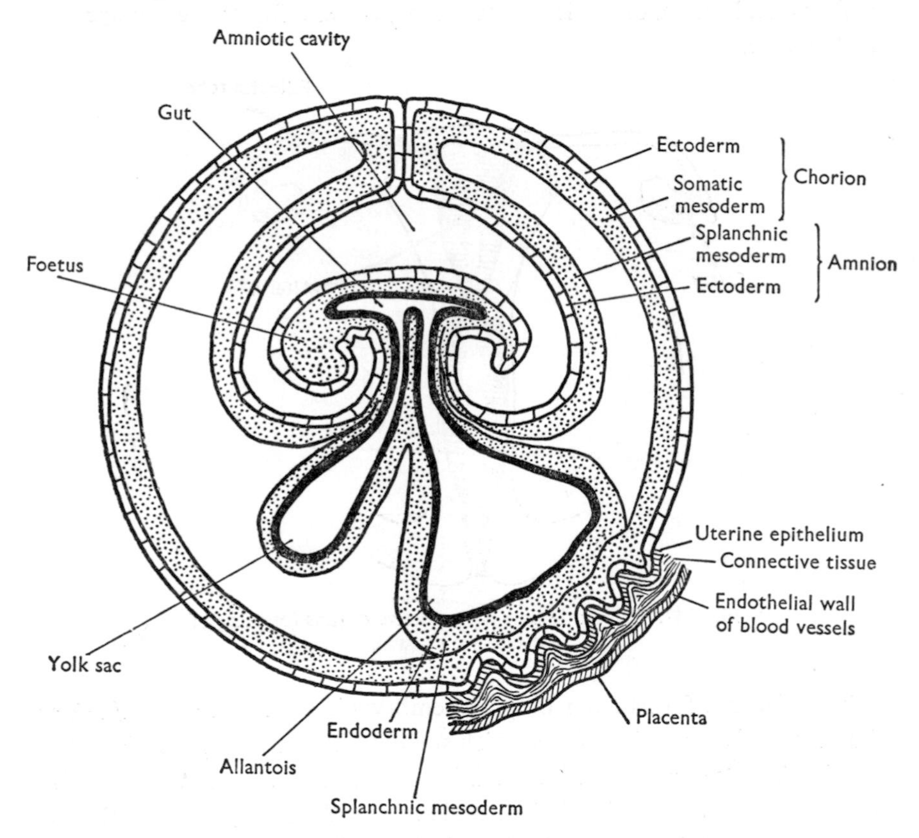

Fig. 148. Diagram of placenta formation in mammals.

(The foetal circulation and the by-passing of the lungs has already been described on p. 91 and the passive immunisation effect of the mother's circulating antibodies is discussed on p. 120.)

The testes of the male mammal usually descend out of the body into the scrotum, which accounts for the looping of the vas deferens round the ureter. The former enters the top of the *urethra*, and, at this point, a diverticulum may form a *seminal vesicle* and *coagulating gland.* Other accessory male glands are the *prostate* and *Cowper's glands* and these produce hormones and nutritive substances which are added to the sperm to make up the seminal fluid. The copulatory organ for internal fertilisation is the *penis* whose vascular *corpora cavernosa* and *spongiosum* allow it to become erect during mating.

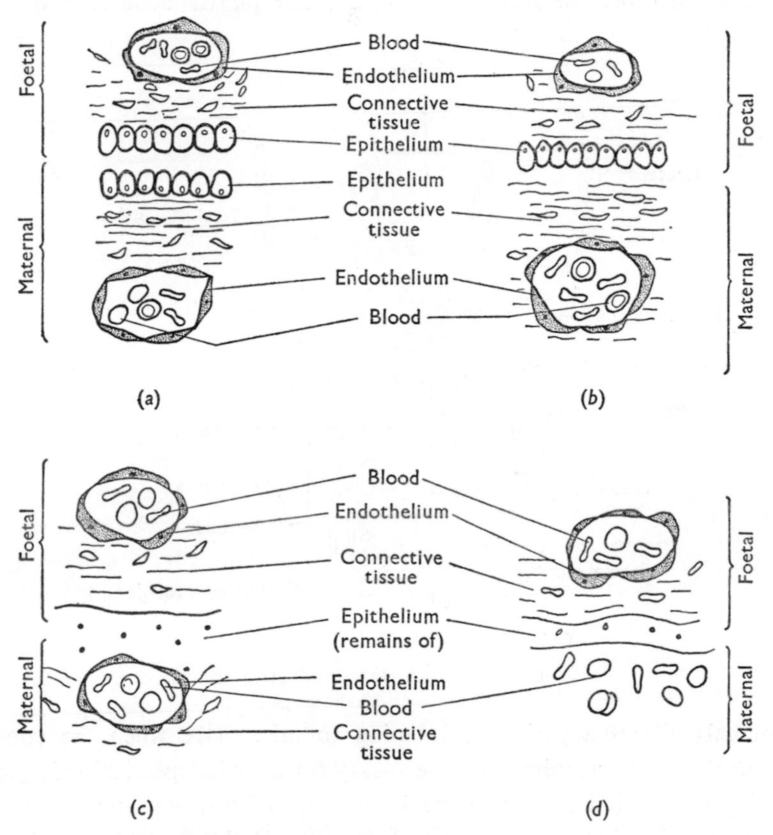

Fig. 149. Types of placenta. (*a*) Epitheliochorial (horse); (*b*) syndesmochorial (cow); (*c*) endotheliochorial (dog); (*d*) haemochorial (man).

Besides the genitalia, development of vivipary and the placenta, which have already been described, the reproductive efficiency of mammals has been greatly increased by post-natal care of the offspring. Glands, derived initially from sebaceous glands, are developed in the female of the higher mammals, though not in the monotremes; these are concentrated as mammary glands *anastomosing* (running together) into a surface nipple. The glands enlarge during gestation and milk is secreted shortly after birth. The composition of this varies somewhat with the animal concerned and its speed of growth, but all mammals produce a highly nutritive and essentially balanced food-substance upon which the offspring can thrive. Coupled with this feeding is the intense development of parental care, especially in the female, directed towards the survival of the young. Besides the direct defence of the latter, the plastic behaviour of the

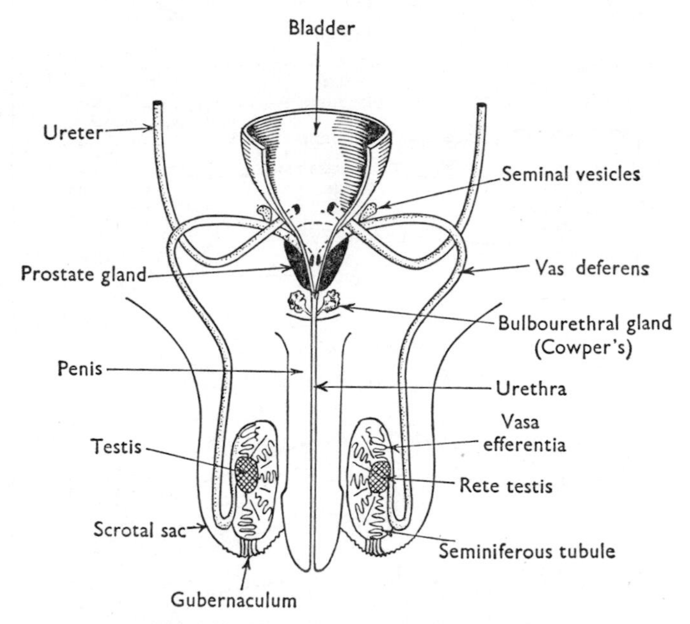

Fig. 150. Human male reproductive organs.

mammals allows a period of learning to take place when the young mammal actually acquires skills necessary for later independent existence. During this period there is a maximum rate of learning and degree of communication between parent and young and there can be no doubt that this important phase of life has been very greatly extended in our own species.

The male has some role to play in this rearing of the young but his activity is supplementary and in some species the male tends to disperse his own young away from the 'nest' as they begin to mature.

11 THE CELL

11.1 Introduction—the concept of the cell

Most of the processes that have been described in the foregoing chapters can ultimately be related to activities taking place at a cellular level. With the higher magnifications that can now be achieved with the electron microscope and the increasing delicacy of research techniques in biochemistry and microbiology, many of these physiological processes are being traced not only to their cytological but also their molecular basis within the protoplasm.

Nevertheless, and in spite of the great advances that have been made in recent years, it is well to remember that as with other rapidly advancing subjects it is not always easy to distinguish between established fact and attractive hypotheses. The state of our understanding of the genetic code and of protein synthesis are clear examples of this situation in the early 'sixties. It is also true of many other fields of functional investigation of living organisms and it must never be assumed that the most recent views are necessarily the nearest to the truth. One of the difficulties of modern techniques for the investigation of protoplasm is their dependence on studies *in vitro* of extracts broken down from the living organism. In other cases there is an isolation of particular functional units of the cell such as the *microsomes, nucleic acids* or *mitochondria.* Sometimes the nature of the media essential for such *in vitro* experiments is far from that which naturally occurs. There is a danger therefore that our understanding of the integral activity whereby the whole cell 'lives' will become compartmentalised and may be artificial. It is from the 'getting together' of biochemistry, cytochemistry and electron-microscopy that the more realistic advances in our knowledge of the cell are being built up. It is to be hoped there will be eventually a synthesis of these studies at lower levels which will become integrated with those studied at the physiological level.

Cells were observed as far back as the seventeenth century by Van Leeuwenhoek but it was Schleiden and Schwann who in 1839 brought out a theory of the cellular organisation of living matter. The cell theory has been a useful working hypothesis for biologists and it is certainly correct to think of animals as being divided up into these protoplasmic units, each controlled by a single nucleus. At the same time it must be remembered not only that there may be protoplasmic connexions between the cells, but also that organisms work as continuous systems for, although all cells have the same genetic potentialities, each contributes in a special way to the life of the whole.

We shall consider the cell membrane, the exchange of energy within the cell and the synthesis of protein and genetic materials, as these are subjects which have arisen in other chapters and about which much recent knowledge exists. However, these are certainly not all the activities of cells, and they tell us little of the problems of diversity and specialisation of individual cells. Although a great deal is known about the latter subject it is not easily summarised and much of it is outside the scope of the present account.

11.2 General anatomy of the cell

A generalised animal cell may be thought of as being roughly spherical and some 5μ ($\frac{5}{1000}$ mm) in diameter, so that a large multicellular creature like man may contain some 10×10^{14} cells. Each cell has a semi-permeable covering called the *plasmalemma* (or plasma membrane) which shows selective absorption and other properties described below. The plasma membrane surrounds the cytoplasm—that is, the general mass of protoplasm which contains large organelles which are readily seen with a light-microscope and whose complex organisation has been demonstrated at higher magnification. The larger inclusions are rectangular *mitochondria* (energy exchange), *lysosomes* (autolysis), fat droplets, *Golgi apparatus*, *centrosomes*, etc., and the more detailed organisation consists of the *endoplasmic reticulum* and the *ribosomes* which are associated with protein synthesis. Specialised cells also have drops of secretory material, pigment granules and colour organelles.

From chemical analysis cytoplasm has been shown to consist of the following: water 80–90 %, proteins 7–10 %, lipoid 1–2 %, carbohydrate 1–1·5 %. The substances are not in true solution but exist in a colloidal form—or rather the special liquid-in-liquid state called an emulsion. This means that many of the larger molecules, which are called the *disperse phase*, are in a permanent state of suspension within the liquid, which is mainly water and ions and constitutes the *continuous phase*.

Colloids have a number of properties which are of interest in the working of biological systems, chief of which is the great surface area of the disperse and continuous phases which allows the permanent availability of many chemical templates at which reactions may occur. The colloid is unstable and can be precipitated out if conditions change; it is particularly sensitive to temperature, acidity, and degree of hydration. The colloid also has a viscosity which can change and is important in some of the streaming

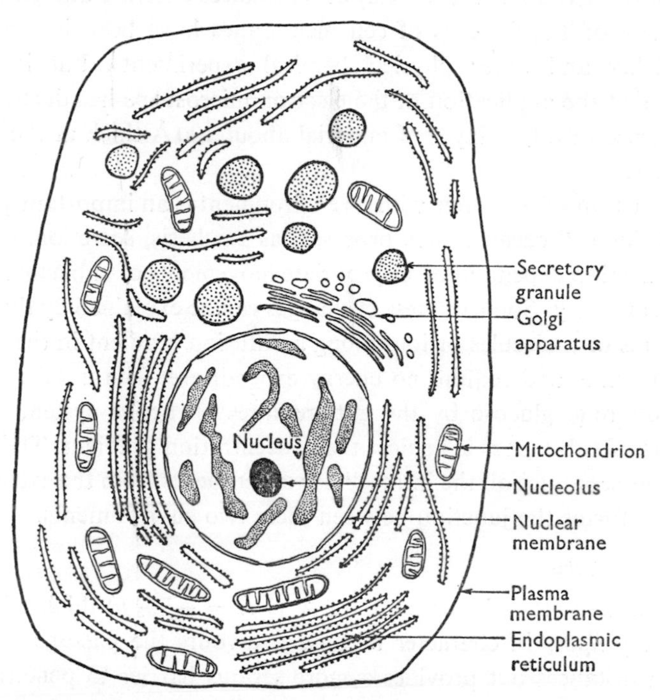

Fig. 151. Diagram of a secretory cell from the pancreas, with well-developed endoplasmic reticulum and Golgi apparatus. All of the structures shown, with the exception of the secretory granules, are found in almost all cells, but their abundance and relative proportions vary widely from one kind of cell to another.

activity of protoplasm as well as in the formation of surface membranes. Radicals, ions and molecules within the colloid will exert an osmotic force across the semi-permeable membrane of the plasmalemma, drawing water into the cell.

Within the mass of cytoplasm is the nucleus, this again having a surrounding membrane (nuclear membrane) and containing the *nucleolus* with its store of nucleic acid in addition to the chromosomes with their genetic materials. The functions of the nucleus in the control of the cell and reproduction will be dealt with after we have considered the cell membranes.

11.3 The cell membrane

Although the whole concept of a specialised membrane surrounding the cytoplasm of cells has been questioned many times, a great deal of work has been done in the last twenty years to show the nature of the cell surface and how it operates in the control of substances entering and leaving the cell. Many of the features of cell membranes have been inferred from permeability and other physico-chemical experiments, but it is only recently that the application of the electron microscope has demonstrated the existence of a thin layer of material about 100 Å thick at the borders of all cells.

The function of exchange with its environment is an important property of the living cell because such processes as synthesis, secretion, excretion and conduction depend upon appropriate movements of substances across the cell boundary. Some of these exchanges can be explained by the passive movements of molecules or ions along a diffusion gradient in the pores of the membrane, and require no energy expenditure. The uptake of other substances (e.g. glucose by the gut) requires an active system, because transport of substances is against the concentration gradient. The plasma membrane separates off the intracellular constituents from the extracellular fluid, and forms the junction between these two environments.

11.31 *Structure*

The cell membrane is some 80–100 Å (1 Å = $\frac{1}{10,000}$ μ) thick. It is predominantly *lipoid* in character because it favours the passage of lipoid-soluble substances but provides a more serious barrier to penetration of water and water-soluble substances. It probably consists of a sandwich of lipoid material with the polar groups pointing outwards and they are stabilised by a thin layer of protein. The protein is interspersed with phosphoric acid. Scattered throughout this membrane are occasional pores, protein lined, which allow the rapid entry of certain substances (e.g. glycerol). The thin protein layers account for the viscosity and elasticity of the membrane as well as its surface tension effects, while the fatty layer makes it possible to understand how it is readily permeable to fat-soluble substances. Finally the pores account for the semi-permeable effects of the membrane to certain sizes of molecules (Fig. 152).

Both inner and outer surfaces of the membrane are richly charged with organic groups and ions and many of the properties of the cell are determined by the configuration of its membrane surface. The way in which antibodies react with certain specific surface arrangements of their antigens has already been described (p. 116).

Besides their local variations of structure, cell membranes may also be charged at their surface because of differences in the distribution of charged particles on the two sides of the membrane, and this is important in conduction and muscular contraction.

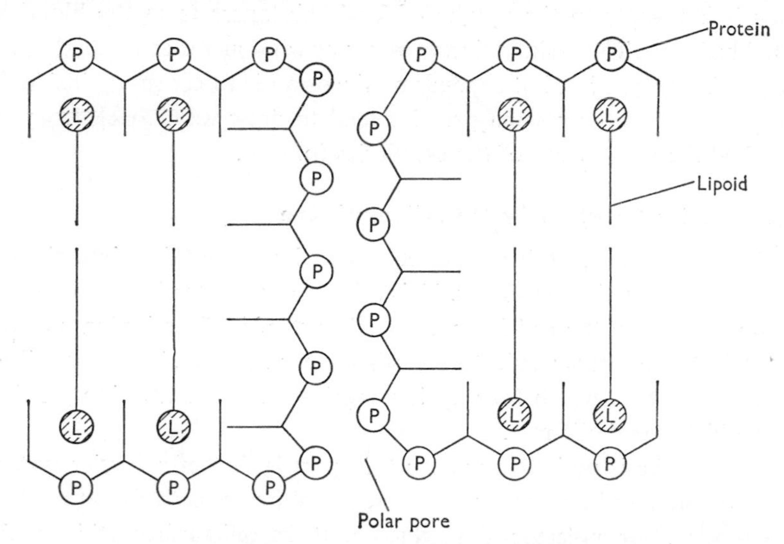

Fig. 152. Arrangements of the protein and lipoid components of the cell membrane.

11.32 *Passive uptake and loss of substances through the cell membrane*

At rates which follow from the laws of diffusion, substances move from regions of high concentration to those of low concentration and, where the membrane is readily permeable, such a process will naturally occur between the *intra-* and *extracellular fluids.* Should the substances concerned become bound once they are inside the cell by their incorporation into large molecules, they will no longer have any effect on the diffusion gradient, and the substances will continue to pass across the membrane in a passive way, although the forces necessary to bring about the binding of cell constituents may require energy.

If molecules exist in larger concentration on one side of the membrane they will cause osmotic effects whereby water will be drawn from the hypotonic to the hypertonic medium. Such an effect can be seen when red blood cells are placed in distilled water, when they swell up and burst, or, when they are transferred to strong salt solution, water passes in the opposite direction and they decrease in volume. This is also the basis of the mechanism whereby cells at the end of the kidney tubules reabsorb water from the tubular fluid and so leave a hypertonic urine (see p. 132).

Such passive movements apply to all substances which can pass across the cell membranes or through its pores, and where a diffusion gradient exists such that the concentration outside the cell is greater than inside. Thus it can apply to the uptake of food substances from the blood by individual cells. The assimilation of water by cells, as for example those of the gut, is also passive, but depends on the osmotic forces set up inside the cytoplasm—thus the plasma proteins tend to draw water back from the lymph at the venous end of the capillaries (p. 106).

11.33 *Active transport by the cell membrane*

Many substances, both as molecules and ions, are taken into the cell against concentration gradients. This selective concentration of necessary substances within the cell is an essential part of its functioning and can only take place where respiratory mechanisms are intact. Thus the cells of the kidney tubule which take up ions against the concentration gradient have a high metabolic rate.

One hypothesis underlying active transport is that of a carrier system. It can be supposed that such a substance combines selectively with certain ions (e.g. K^+) or molecules (e.g. glucose) at the cell surface—this combination requiring metabolic energy—and that the new complex is now able to pass back across the membrane and its inner surface releases the ions or molecules it has been carrying. The carrier is then reactivated. This idea has been extended further to suppose that the reactivated carrier may now combine with sodium ions and on passing to the surface releases them, thus forming the basis of the *sodium pump*, common to animal cells. Different substances transported would require different carriers, but all would require energy for their activation, which would account for the high metabolic rate of such tissues.

It must be realised that such mechanisms are as yet hypothetical and have not been shown to work during active transport.

Since cells have been observed by cine techniques it has become realised that other phenomena operate at their surface—thus the membrane is exceedingly active during the life of a cell, continuously pushing out fingers of protoplasm into the surrounding medium and sucking in little vacuoles of fluid. These invaginations are seen to shrink and divide and there is no doubt that their contents are being taken across the cell membrane. This has given rise to the idea that cells drink in necessary substances—hence the phenomenon is called *pinocytosis.* From the point of both active and passive transport there can be no doubt that this activity greatly increases the surface of the cell membrane in contact with the extracellular fluid and thus assists movements across it. Phagocytosis, which is a sort of gross assimilation of foreign substances into the cell, also takes place by invagination of small areas a little at a time, and their gradual solution into the cytoplasm.

11.4 Energy exchange

11.41 *Role of energy carriers*

One of the fundamental properties of living matter is its constant demand and utilisation of energy, which it requires for obtaining its organisation and for the performance of its activities. The large molecules at whose surfaces reactions take place, can only be built up and function

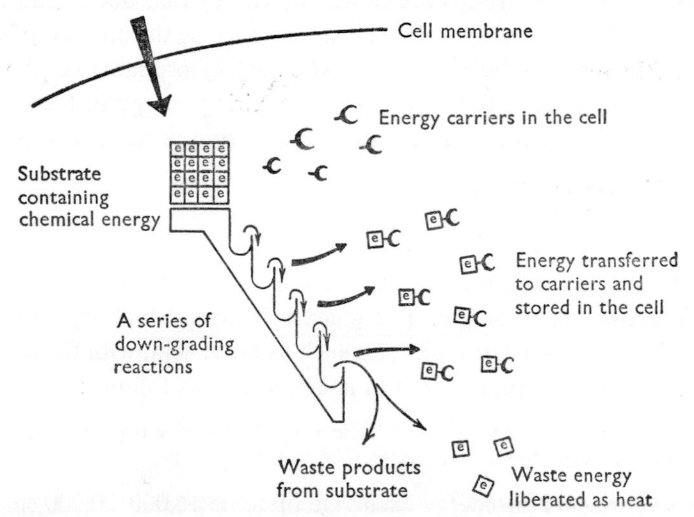

Fig. 153. Energy exchange in the cell.

where energy is available. In the cell, energy is mainly carried as a substance called adenosine triphosphate (ATP), which acts as the immediate energy provider in all animal cells.

Before seeing how this substance is formed and operates in the cells, it would be as well to consider the general problem of energy pathways. The basic factor in which living systems differ from non-living energy-exchange systems (such as engines) is that the organic components of the former work at lower energy levels. Living systems use small quantities of energy appropriately made available and applied as required. If all the energy of the fuel substrates used by the cell was liberated at once, most would be lost as heat. In order to reduce such losses and to function at lower temperatures, energy exchange is accomplished by a series of downgrading reactions whereby energy is released in small packets. This sort of idea can be expressed diagrammatically as shown in Fig. 153.

The type of substance that can act as an energy store or carrier must have a molecule which is stable yet from which parts can be separated readily with the release of energy. Such molecules frequently have specific energy-rich bonds which are associated with various organic phosphates, the most common being part of the adenosine series. This substance *adenosine* consists of a purine base (double ringed with four nitrogens linked to five carbons) linked to a five-carbon (ribose) sugar which can be phosphorylated up to three times to form mono-, di- and triphosphates. The second and third phosphate groups are linked by energy-rich bonds and release 8000–10,000 cal/mol on hydrolysis. This seems to be the amount of energy which can be readily utilised in the cell and the triphosphate to diphosphate bond is normally used for all purposes requiring energy in the cell. The series of reactions involved in the phosphorylation of adenosine is shown on p. 265.

11.42 *The first stage of energy release (glycolysis)*

The energy exchanges between glucose[1] and carriers in the cell take place in a number of stages. The glucose is first phosphorylated and is broken down into smaller molecules such as lactic acid, with the release of energy. This may be represented in a simple way as follows:

Glucose + 2 adenosine diphosphate → 2 lactic acid + 2 adenosine triphosphate

$$C_6H_{12}O_6 + 2ADP + 2PO_4 \rightarrow 2CH_3CHOH.COOH + 2ATP$$

and this represents an energy exchange of some 16,000–20,000 cal/g mol (i.e. per 180 g of glucose). This is only a small amount of the total energy available in the glucose (690,000 cal/g mol) but takes place in the absence

[1] The glucose may have been recently taken across the cell membrane or have been derived from stored products such as glycogen within the cytoplasm, which needs to be prepared for oxidation by a decrease in the size of their molecules by processes similar to those of digestion. These enzymes are liberated by the *lysosomes* of the cell. (Lysosomes are small circular inclusions observable in protoplasm which contain and in a sense restrict the activities of digestive enzymes—their breakdown leads to *autolysis* or self-digestion of the cell.)

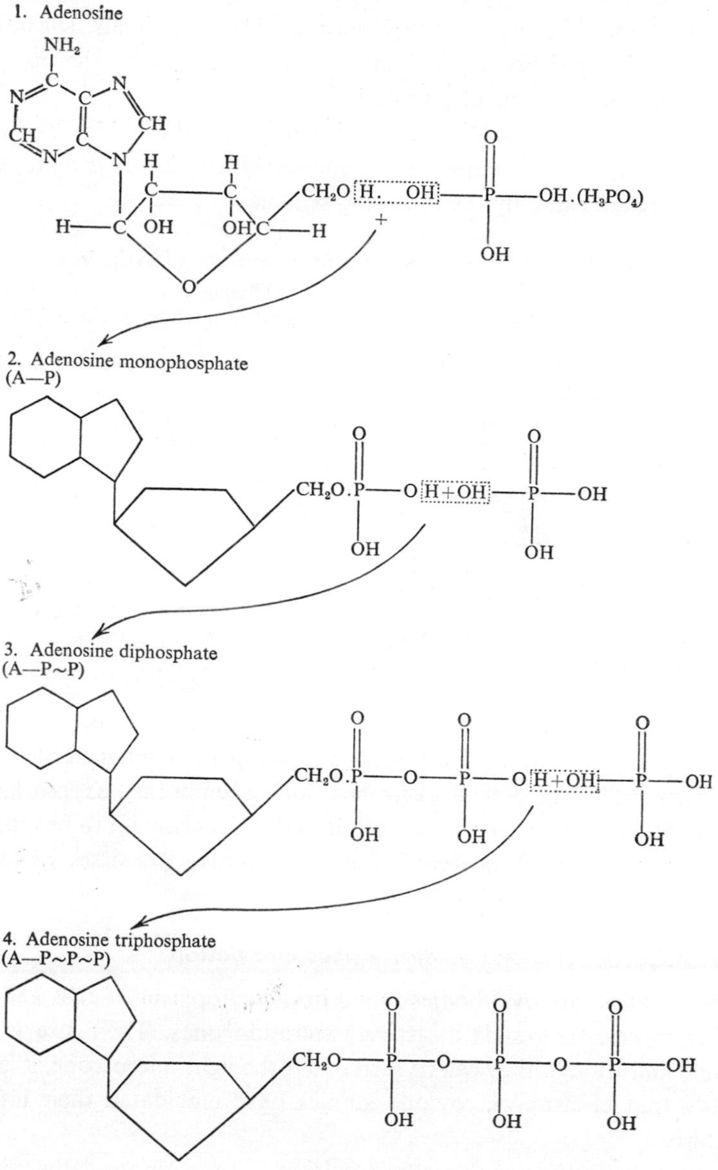

1. Adenosine

NH_2 N C C N CH CH N C N H C H C CH_2O H. OH — P — OH.(H_3PO_4) O OH H—C OH OH C—H O +

2. Adenosine monophosphate
(A—P)

CH_2O.P—O H+OH —P—OH O OH O OH

3. Adenosine diphosphate
(A—P~P)

CH_2O.P—O—P—O H+OH—P—OH O OH O OH O OH

4. Adenosine triphosphate
(A—P~P~P)

CH_2O—P—O—P—O—P—OH O OH O OH O OH

(See opposite.)

of oxygen. Such anaerobic respiration is used as the only form of exchange by some animals (e.g. a few internal and gut parasites), certain bacteria, and from time to time by yeast and higher plants. It is very inefficient and insufficient to supply the energy requirements of larger animals. The normal way of supplying these requirements is only achieved by the complete oxidation of the products of glycolysis.

The detailed pathway of this process is complex and considerable knowledge of biochemistry is required to understand it fully. It is represented in an abbreviated way by the following scheme:

Table 15. *Energy changes in anaerobic glycolysis*

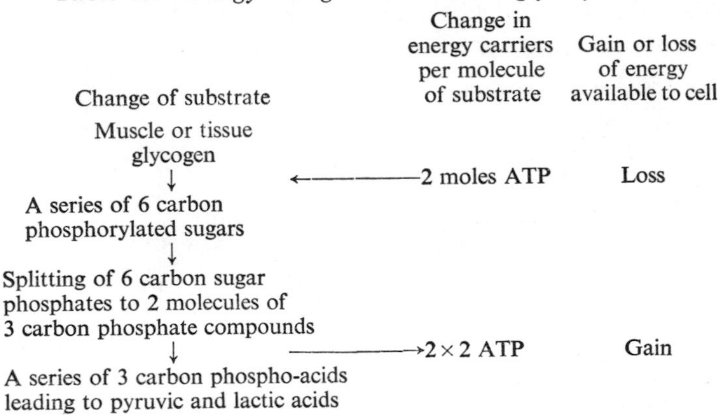

Change of substrate	Change in energy carriers per molecule of substrate	Gain or loss of energy available to cell
Muscle or tissue glycogen ↓	← 2 moles ATP	Loss
A series of 6 carbon phosphorylated sugars ↓		
Splitting of 6 carbon sugar phosphates to 2 molecules of 3 carbon phosphate compounds ↓	→ 2 × 2 ATP	Gain
A series of 3 carbon phospho-acids leading to pyruvic and lactic acids		

This is the end of glycolysis and so far no oxygen has been used—or rather—as the oxidation of $DPNH_2$ can take place later it need not yet have been used. The lactic acid formed during temporary oxygen lack in the cells accumulates and can pass out into the bloodstream to be returned to the liver or else it can enter further oxidative processes which are associated with the mitochondria.

11.43 *Mitochondria and oxidative phosphorylation*

Mitochondria are oval bodies found in the cytoplasm of cells and may number several thousands in actively secreting ones. Their size is very variable and though they can be seen under the light microscope, it is only recently that electron-microscope studies have elucidated their internal structure.

Around the outside of the mitochondrion is a membrane rather similar to that of the plasmalemma—that is, two layers of lipoid sandwiched between protein layers, and doubtless with the same sort of function of selective uptake of substrates metabolised by enzymes inside the mitochondrion. The inner surface of the mitochondrion is thrown into numerous folds or *cristae* bathed in a fluid matrix within its vacuole. It is on this inner surface that the sequence of reactions involved in the aerobic oxidation of pyruvate produced by glycolysis occur.

There are two general stages in this oxidation: the first is by means of the *Krebs, tricarboxylic* or *citric acid cycle* whereby pyruvates are broken down to CO_2 and 'H' (in the form of a reduced hydrogen carrier) and the second entails the passing on of this hydrogen to molecular oxygen, thus

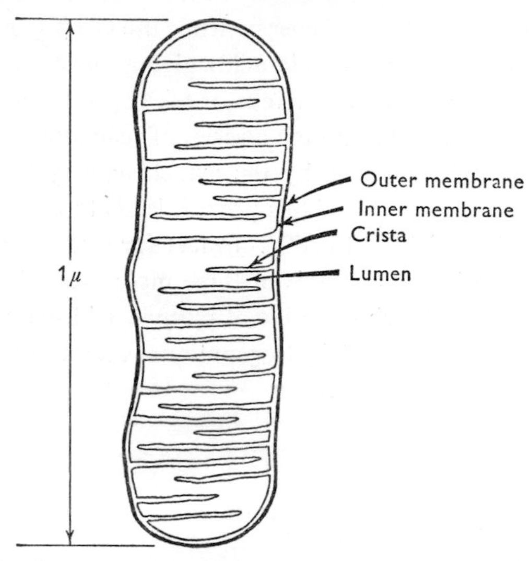

Fig. 154. Mitochondrion from pancreas cell as seen in electron micrograph. Cristae are formed as infoldings of the inner membrane; they are the site of the cytochrome system. Enzymes of the citric-acid cycle occur in the lumen of the mitochondrion.

forming water. Both of these stages release large amounts of energy which becomes stored as ATP and other carriers such as creatine phosphate. Altogether there are some twenty-five sets of enzymes involved in the complete oxidation of pyruvic acid and the size of the mitochondrion allows these individual sets to be represented some two thousand times. By this oxidation there are about 300,000 cal transferred into high-energy phosphate per g mol of glucose, which can either remain in the mitochondrion or pass out into the cell, where it is used for a variety of purposes. Thus the mitochondrion can quite reasonably be thought of as the 'power house' or engine of the cell, and those cells that do most work and have the highest respiratory rate (e.g. liver) are the richest in mitochondria (see Fig. 154). It is also interesting that the structure of mitochondria and chloroplasts is extremely similar as the latter of course carry out another type of energy-exchange involved in photosynthesis.

11.44 *The citric acid cycle*

Pyruvic acid from glycolysis is produced in the cytoplasm and enters the mitochondrion where it acts with coenzyme A to produce acetyl CoA:

$$CH_3CO.COOH + \text{Coenzyme A} \rightarrow CH_3CO\text{-Coenzyme A} + CO_2 + 2H.$$

CO_2 is released and diffuses out of the cell, the 2H reduces the hydrogen carrier DPN (diphosphopyridine nucleotide) to $DPNH_2$ which is later oxidised with the release of energy, while the CH_3CO group combines with another molecule (oxaloacetic acid) giving citric acid which is the first substance of the citric acid, Krebs or tricarboxylic acid cycle.

This citric acid goes through a series of reactions some of which release hydrogen that is taken up by the cytochrome system and subsequently combined with oxygen, and this process leads to the release of energy and the conversion of ADP to ATP. In fact for each pair of hydrogen atoms released there is the synthesis of three molecules of ATP. For the complete oxidation of each molecule of pyruvic acid there are three molecules of CO_2 formed and five dehydrogenations (via DPN or TPN or direct to a flavin-linked hydrogenase).

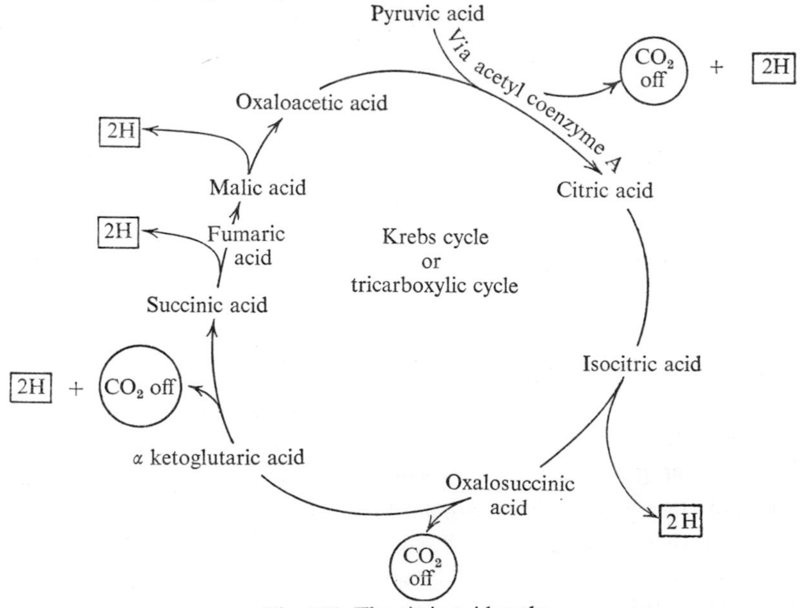

Fig. 155. The citric acid cycle.

11.45 *Cytochrome*

As we have indicated, the last and most vigorous release of energy takes place in the transfer of the H from the DPN formed in glycolysis and the Krebs cycle to molecular oxygen. This takes place through a number of intermediate transfer reactions, some of which release energy for the phosphorylation of ATP.

Cytochromes are iron-containing chromoproteins which have the haem prosthetic group linked to proteins with molecular weights of about 16,000. There are several components in this system and the most important are cytochrome C and cytochrome A (cytochome oxidases). The latter enzyme is inhibited by *cyanide* and this is why 80–90 % of tissue respiration is abolished by this poison and with such disastrous results. The role of these carriers in the receipt of hydrogen from the breakdown products of glucose is as follows:

AH_2 → oxidised ← → H_2O *
Cytochrome
A ← → reduced ← $\frac{1}{2}O_2$

In fact the substance labelled AH_2 is a more complex molecule (flavoprotein) which has received its hydrogen in turn from a nucleotide (DPN or TPN) which in its turn has received it from an organic molecule resulting from the degradation of glucose. The chain of reactions is as follows:

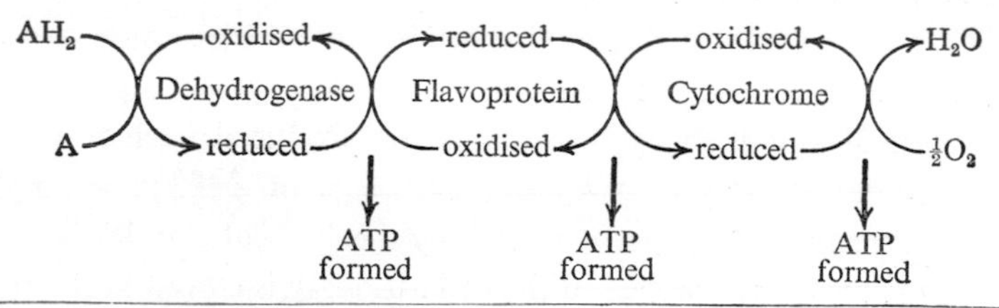

---

* 2H from hydrogen carrier

H, H

Proton + electron | Proton + electron

Cytochrome oxidase

2 Protons | $\frac{1}{2}O_2$

$2Fe^{+++}$ Cytochrome c $2Fe^{++}$

2 Electron

H_2O

Actual transfer of the H atom across cytochrome.

During these reactions, therefore, three molecules of ATP are produced for each pair of hydrogen atoms that are removed from a given substrate in the glycolysis or citric-acid cycles. In Fig. 156 the computation of the total number of ATP molecules produced during the degradation of a single molecule of glucose is indicated and it is clear that 30 molecules of ATP are produced during the citric-acid cycle. Under aerobic conditions there are also six ATP molecules formed aerobically in the transfer of hydrogen from the $DPNH_2$ formed during the oxidation of triosephosphates of glycolysis. In addition to this total of 36 ATP we have the 2 molecules of ATP which are the net gain of anaerobic glycolysis. It is apparent therefore that during aerobic respiration there are nineteen times more molecules of ATP synthesised than during anaerobic respiration.

11.46 *Summary on energy exchange*

In order to make the above account clearer we may consider a number of questions about the energy exchange in the cell and answer them from what has been described.

The overall breakdown of glucose to CO_2 and water may be indicated by the equation

$$C_6H_{12}O_6 + 6O_2 \rightarrow 6CO_2 + 6H_2O + 690 \text{ Cal}$$

when the glucose is oxidised outside the body by normal combustion. The questions we may ask are:

(1) What proportion of this available 690,000 calories contained in the glucose molecule is released in a form utilisable by the organism? The answer is that 38 mol of energy-rich carrier are produced, each of which can yield 10,000 cal—thus giving a total exchange of $\frac{380000}{690000} = 55\%$ efficiency; a very satisfactory performance by mechanical standards.

(2) How is the energy stored in the cell? Energy is stored in the form of adenosine triphosphate (ATP) and creatine phosphate (CP) high-energy bond substances whose terminal phosphate releases 10,000 cal on hydrolysis.

(3) For what purpose is the energy stored? The energy of the ATP can be made available to form links between the muscle proteins, thus causing contraction and performing mechanical work. It can be released in the form of heat energy (this is a by-product of other uses). It can be used to activate amino acids and other chemicals of the protoplasm, causing them to be condensed into larger molecules for the synthesis of further protoplasm, secretions, etc. It can also be used in the production of light and the active transport of substances across membranes.

(4) Where does the oxygen get involved in the cellular respiration, and where is the CO_2 released? As we have seen, the molecular oxygen plays a vital part, as it is the final hydrogen acceptor of the cytochrome system through which 90 % of the tissue respiration proceeds. In the absence of this part of the cellular mechanism only one-nineteenth of the energy-rich phosphate can be made available as ATP. The CO_2 is all released in the citric-acid cycle.

(5) How can the changes undergone by the respiratory substrates be indicated in a simple fashion which indicates the main steps? This is best done by a diagram such as that shown on page 272 (Fig. 156).

So far we have considered the degradation of carbohydrates, which mainly proceeds as a first step through the glycolytic cycle. There are alternative pathways for the degradation of glucose, but they play a relatively small part. The use of fuel substrates other than glucose, however, is an important part of the cellular mechanism which we will now consider.

While the cell usually makes use of glucose to provide its energy, both fat and protein can also be respired to yield energy for ATP formation.

Fats are converted into fatty acids in the cell by hydrolysis and these are first activated by coenzyme A (see citric-acid cycle), a hydrogen molecule being removed at the same time and transferred to a flavoprotein subsequently oxidised with energy release. The fatty acid is now cleaved and the CH_3CO (acetyl) group removed by the coenzyme A, which is passed into the beginning of the Krebs cycle and releases its energy as outlined above. This process is repeated until a long chain of fat is broken down into acetyl-acetic acid, $CH_3CO.CH_2COOH$, and finally completely oxidised. It is calculated that a six-carbon fatty acid will yield some 44 ATPs as compared with the 38 of a six-carbon sugar so they can generate rather more energy. The fate of the glycerol component of the fat is to be degraded along one of the alternative carbohydrate pathways. A simple scheme for fat respiration is indicated on page 273 (Fig. 157).

Proteins are normally hydrolysed to their constituent amino acids and these undergo deamination whereby ammonia is removed and a keto-acid formed:

$$R.CH.NH_2.COOH + \tfrac{1}{2}O_2 \rightarrow R.CO.COOH + NH_3.$$

The 2H are removed by flavoprotein as FPH_2 and are passed to the cytochromes. The keto-acid is now converted to acetyl coenzyme A—either directly if it is a long chain one, or by intermediate stages if complex side changes are present. The acetyl coenzyme A is then fed into the Krebs cycle. In some cases the amino acids do not yield acetyl coenzyme A but

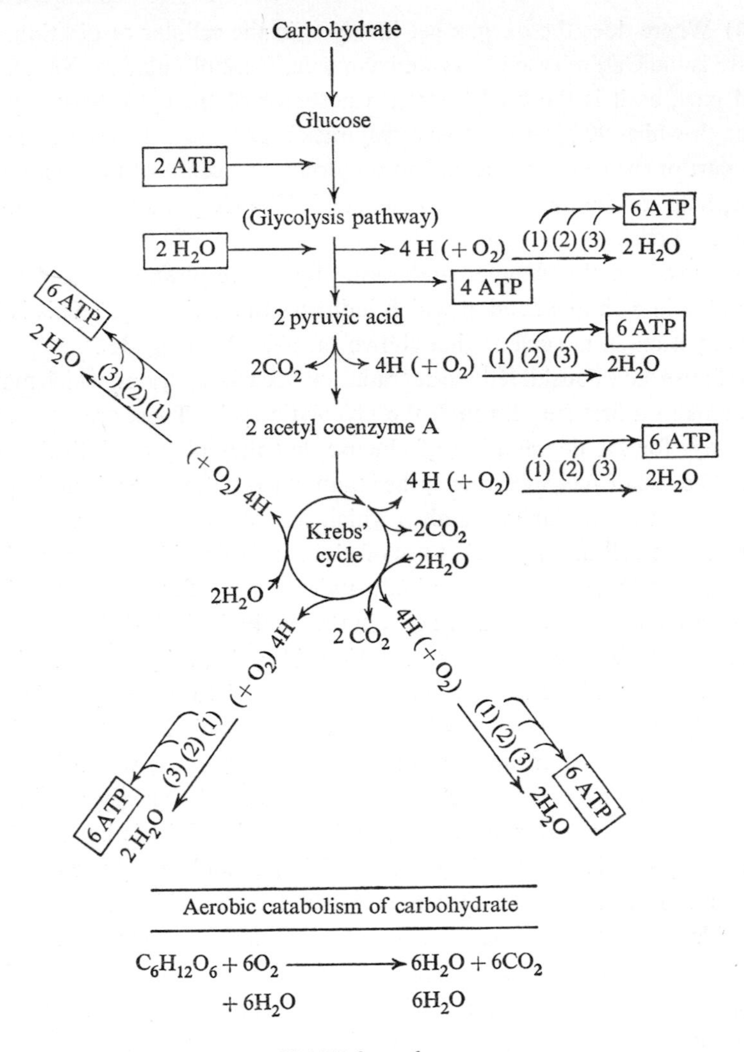

Fig. 156. Changes undergone by respiratory substrates. By inspection, it can be seen that the role of each of the participants in the original reaction summary

$$\text{(i.e. } C_6H_{12}O_6 + 6O_2 + 38ADP + 38PO_4''' \rightarrow 6CO_2 + 6H_2O + 38ATP)$$

is indicated.

pyruvic acid or other intermediate stages of the Krebs' cycle, for example, α-ketoglutaric acid. The oxidation releases most of the original energy available in the amino acids. It should be realised that one amino acid can be transaminated and become converted into another and used in synthesis, or used directly.

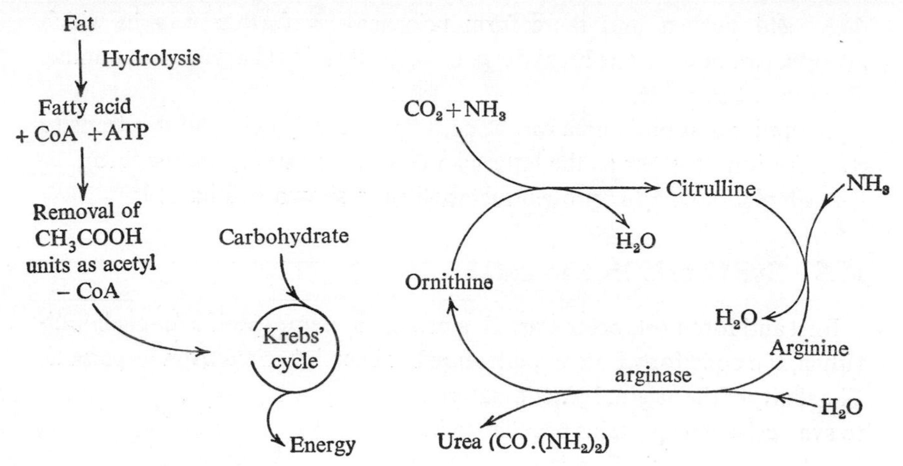

Fig. 157. Fat respiration.

Fig. 158. Ornithine cycle.

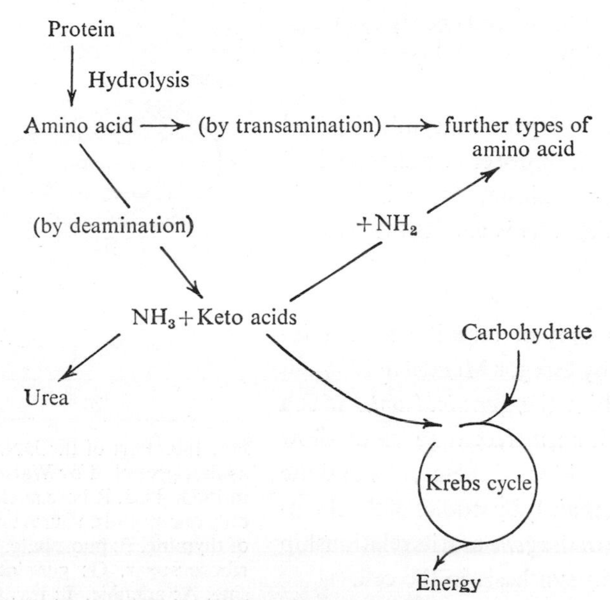

Fig. 159. Protein metabolism.

The NH_3 formed by deamination, in either liver or kidney, is passed into the ornithine cycle (Fig. 158) together with CO_2. As indicated in Fig. 158 urea is formed by the operation of this cycle.

Urea is far less toxic than ammonia and makes up the major part of the nitrogenous excretory matter of mammals—the blood and urine containing only a very small percentage of ammonia. Nucleic acids, taken in the diet, yield purines and these form uric acid—a further way in which nitrogen can be lost. The breakdown of body proteins also yields creatinine, also present in urine.

In reptiles and birds urea can be converted to uric acid and this leads to conservation of water as the latter can be excreted in crystalline form.

A simple scheme for protein metabolism is shown in Fig. 159.

11.5 Synthesis in the cell

First and foremost, protoplasm is a protein system, specific for each individual, the code for whose manufacture has been inherited from its parents. The ability of the original genetic material to synthesise new proteins peculiar to its own type of organisation is one of the characteristic properties of living things. Together, synthesis and energy exchange are the most fundamental properties of the cell.

The basis of protein synthesis, and hence the make-up of enzymes, secretions and new protoplasm, depends initially on the nucleus. This carries the *chromosomes*, which are themselves composed of strings of genes which represent the particulate unit of inheritance first investigated by Gregor Mendel in 1866 and called by him the *germinal unit*. Much has been done in recent years to show interrelationships of these units and the cytoplasm, mainly by studies of the chemical nature of the *gene* and its relationship with protein synthesis in the cell.

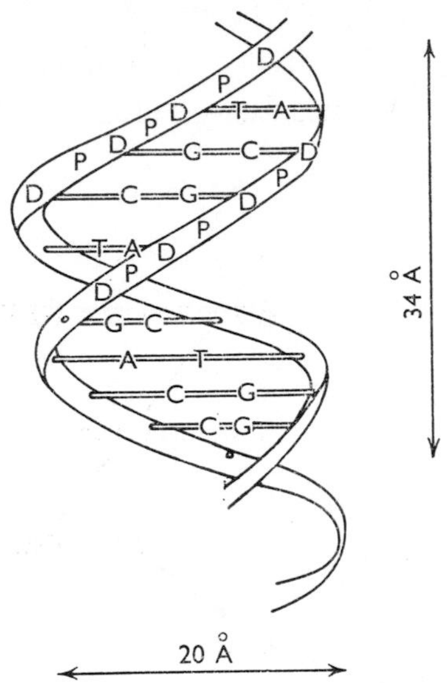

Fig. 160. Part of the DNA molecule as first proposed by Watson & Crick in 1953. N.B. Ribose nucleic acid has only one strand and uses uracil instead of thymine. P: phosphate. D: desoxyribose sugar. G: guanine. C: cytosine. A: adenine. T: thymine.

11.51 *The nucleus*

The main constituent of the nucleus is *desoxyribonucleic acid* (DNA), whose chemical structure was suggested by Crick and Watson in 1957. DNA consists of a double helix of desoxyribose (5-carbon) sugar units alternating with phosphates and bound transversely by four bases in two pairs. These are *adenine* linked with *thymine* and *guanine* linked with *cytosine* (Fig. 160). It is supposed that in the resting cells the DNA genes are actively synthesising chains of nucleic acid which pass out into the cytoplasm, these chains being made up of ribose sugar nucleotides (called RNA). The structure of both DNA and RNA probably act as coded templates on which further molecules of protein may be built up by causing the arrangement and linking together of amino acids in a specific order. The working of such a code requires energy and the cell nucleus cannot function without its cytoplasmic energy-exchange system.

Viruses consist of DNA surrounded by a protein sheath and they have no respiratory mechanism so have to be parasitic on that of their hosts' cells in order to reproduce themselves. Substitution of the DNA structure of certain plant viruses has caused forms to be produced which cannot synthesise their normal enzyme complex.

We must visualise the nucleus of the cell, therefore, as having two functions. During the 'resting' state of the cell the DNAs of the genes are actively forming RNA which passes through the nuclear membrane into the cytoplasm—in other words the nucleus is indirectly determining the nature of the protein synthesised in the cytoplasm. During cell division, however, the nuclear DNA reproduces itself, forming more DNA from nucleotide material in the nucleolus (of the nucleus). This produces the thickening and doubling of the chromosomes during the prophase of mitosis and meiosis and contrasts with their 'lampbrush' appearance during interphase. The replication and exact distribution of sets of the genetic material into the daughter cells in normal cell division (mitosis) is what we would expect of a coding system. This is confirmed by recent demonstrations that certain areas of individual chromosomes are active in different cells during embryological development.

11.52 *Role of the cytoplasm in protein synthesis*

Electron-microscope pictures of the cytoplasm have confirmed the physical measurements suggesting (by such techniques as dropping metal particles through it and orientating iron filings with a magnet) that there is a complex organisation not discernible by light-microscope. This has been termed the *endoplasmic reticulum* and is best thought of as a maze of membranes and enclosed vacuoles running from the cell surface to the nucleus of the protoplasm (see Fig. 151).

Such a reticulum is particularly associated with cells that synthesise quantities of protein such as the liver and is found as the characteristic Nissl granules of nerve cells, whereas it is virtually absent from cells like erythrocytes which do not have a nucleus or much power of synthesis.

It is supposed that the active surfaces for synthesis are represented by particles of RNA or ribosomes observed on the membranes of the reticulum, such messenger RNA having itself been organised by the DNA of the respective gene, whence it moves out of the nucleus. Amino acids taken in across the cell membrane are absorbed on to the RNA surface and arranged into the proteins specific for the cell. This takes place in several stages and requires energy.

First the amino acids become attached to transfer RNAs, which have a special helical structure and are specific for each amino acid. This process

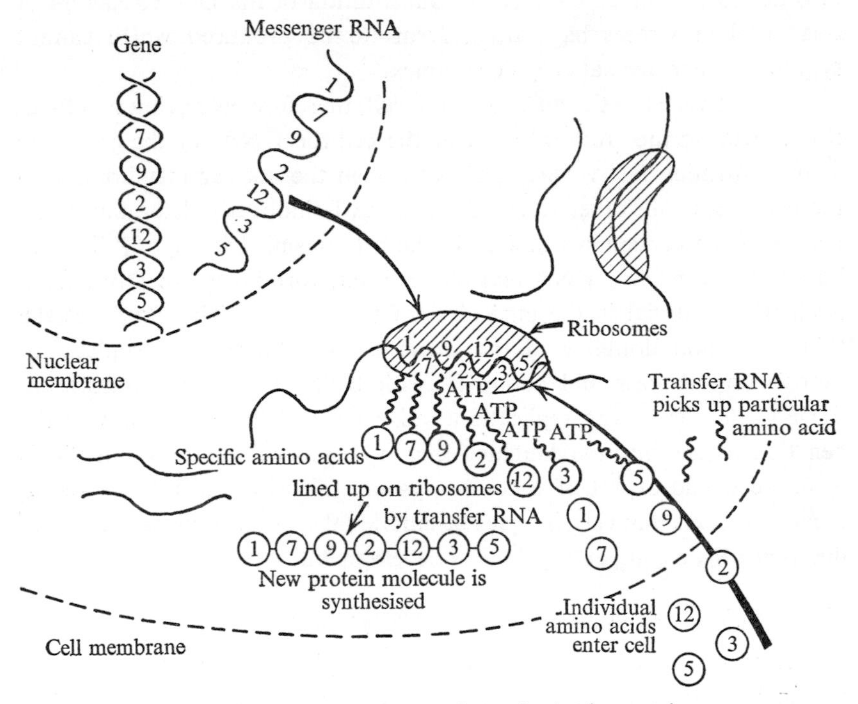

Fig. 161. Sequence of events in synthesis of a protein molecule in a cell.

requires a special activating enzyme and energy derived from ATP. Each transfer RNA is of small molecular weight and has groups specific not only for a particular amino acid but also for its place on the template (messenger) RNA of the ribosomes. Thus the messenger RNA has two sorts of specificity—at one end it picks up the particular amino acid which it is specialised to carry while at the other a code of three nucleotides exists which fits into a three-nucleotide socket on the template RNA. There are some 20 amino acids involved in all protein structures and it has been shown that DNA and RNA probably use a three-nucleotide code having three transverse links for one particular amino acid.[1] As there are three links involved and four different bases the total number of possible combinants is 64 ($4 \times 4 \times 4$). This is more than enough to code for the 20 amino acids and some of these will be represented by more than one sequence of three bases. After the transfer of RNAs have taken up their position on the messenger RNA template changes take place so that the $—NH_2$ and —COOH groups of neighbouring amino acids become condensed, with the loss of water, forming the characteristic peptide linkage —CO.NH— of proteins. Finally the linked amino acids are released from the ribosomes—or may be passed across into the lumen of the ribosome membranes—as proteins carrying out particular functions in the life of the cell or organism.

All these reactions can take place rapidly and it has been estimated that bone marrow cells can combine some 150 amino acids into a molecule of haemoglobin in 80 sec.

[1] Recent work in his field has not only confirmed the triple code but has been able to show the actual code used for specific amino acids. Thus *glutamine* is represented by adenosine–cytosine–adenosine and *alanine* by cytosine–cytosine–guanine on the template RNA.

REFERENCES

General. It is assumed that the students to whom this book is directed will have a background knowledge of chemistry or will be engaged in the study of chemistry concurrent with their biology course.

Certain of the biochemical data required for an understanding of physiology are outside the scope of the normal A-level chemistry syllabuses and it is suggested that the following should be consulted:

Introduction to Biochemistry K. Harrison, Cambridge.

Introduction to Comparative Biochemistry E. Baldwin, Cambridge.

Energy, Life and Animal Organisation J. A. Riegal, English Universities Press.

The following general text-books are relevant to students:

The Life of Mammals J. Z. Young, Oxford.

The Life of Vertebrates J. Z. Young, Oxford.

Structure and Habit of Vertebrate Evolution G. S. Carter, Blackwell.

The Vertebrate Body A. S. Romer, Saunders.

Animal Physiology K. Schmidt-Nielsen, Prentice Hall

General and Comparative Physiology W. S. Hoar, Prentice Hall.

Textbook of Physiology and Biochemistry Bell, Davidson and Scarborough, Livingstone.

For special topics the following are recommended.

Nutrition:

Feeding, Digestion and Assimilation in Animals Jennings, Pergamon.

Respiration:

Comparative Physiology of Vertebrate Respiration G. M. Hughes, Heinemann.

Excretion:

Animal Body Fluids and their Regulation A. P. M. Lockwood, Heinemann.

Osmotic and Ionic Regulation in Animals W. T. W. Potts and G. Parry, Pergamon.

REFERENCES

Movement:

How Animals Move J. Gray, Cambridge.

Bird Flight K. Simkiss, Hutchinson

The Nervous System:

The Conduction of the Nervous Impulse A. L. Hodgkin, Liverpool.

Nerve, Muscle and Synapse B. Katz, McGraw Hill.

The Organization of the Central Nervous System C. V. Brewer, Heinemann.

Endocrinology:

Animal Hormones P. M. Jenkin, Pergamon.

An Introduction to General and Comparative Endocrinology E. J. W. Barrington, Oxford.

Reproduction:

Morphogenesis of the Vertebrates T. W. Torrey, Wiley.

The Sexual Cycles of Vertebrates J. F. D. Frazer, Hutchinson.

The Cell:

The Cell C. P. Swanson, Prentice Hall.

Cells and Cell Structure E. H. Mercer, Hutchinson.

The Life of the Cell J. A. V. Butler, Allen and Unwin.

The Mitochondrion A. L. Lehninger, Benjamin.

The Experimental Basis of Modern Biology J. A. Ramsay, Cambridge.

The Thread of Life J. Kendrew, Bell.

An excellent histology book is:

An Atlas of Histology W. H. Freeman and B. Bracegirdle, Heinemann.

INDEX

INDEX